Wearable Devices - the Big Wave of Innovation

Edited by Noushin Nasiri

Published in London, United Kingdom

IntechOpen

Supporting open minds since 2005

Wearable Devices - the Big Wave of Innovation
http://dx.doi.org/10.5772/intechopen.77458
Edited by Noushin Nasiri

Contributors
Oytun Sözüdoğru, Nazime Tuncay, Menachem Domb, Guo Hong, Dan Liu, Itthipon Jeerapan, Percy Nohama, Taísa Daiana Da Costa, Maria de Fatima Fernandes Vara, Camila Santos Cristino, Guilherme Nunes Nogueira Neto, Tyene Zoraski Zanella, Mart Min, Hip Kõiv, Ksenija Pesti, Eiko Priidel, Paul Annus, Noushin Nasiri

First published in London, United Kingdom, 2019 by IntechOpen
IntechOpen is the global imprint of INTECHOPEN LIMITED, registered in England and Wales, registration number: 11086078, 7th floor, 10 Lower Thames Street, London, EC3R 6AF, United Kingdom
Printed in Croatia

British Library Cataloguing-in-Publication Data
A catalogue record for this book is available from the British Library

Additional hard and PDF copies can be obtained from orders@intechopen.com

Wearable Devices - the Big Wave of Innovation
Edited by Noushin Nasiri
p. cm.
Print ISBN 978-1-78984-496-2
Online ISBN 978-1-78984-497-9
eBook (PDF) ISBN 978-1-83880-342-1

Meet the editor

Dr Noushin Nasiri is the head of NanoTech Laboratory at the School of Engineering, Macquarie University, Sydney, Australia. Her research team focuses on the multiscale engineering of advanced nanomaterials and devices for medical, energy, and environmental applications. Her research lies at the intersection of science, technology, and engineering, as she works on fabricating fingertip-sized nanostructured sensors capable of measuring UV dosage absorbed by skin to improve sun safety and prevent skin cancer, detecting disease biomarkers through analysing human breath, and monitoring levels of environmentally important gases.

Outside the lab, Dr Nasiri is a professional science communicator and her research has been recognized with several prestigious talks including TEDx Sydney in 2017, TEDx Macquarie University in 2019, and TEDx Bligh Street in 2020.

Contents

Preface

Wearable technologies are networked devices equipped with microchips and sensors, capable of tracking and wirelessly communicating information in real time. The rapid adoption of such devices in the past decade has made them the most attractive innovation in the word of technology. From fitness activity trackers to Google Glass, miniaturized wearable devices have shown great potential to be embedded in various domains including healthcare, robotic systems, prosthetics, visual realities, professional sports, and entertainment and arts. Since their first innovation in the 1960s, a wide variety of wearable devices have been developed. These include single functional sensors such as temperature, pressure, and strain detectors, to multifunctional wearable systems capable of monitoring two or more factors simultaneously.

The wearable technologies market is expected to grow 80% from now until 2021, with revenue increasing from $220 million to approximately $12 billion. This is attributed to the wide variety of services these products can offer potential customers. For instance, wireless headset technologies, such as Elinka, OldShark, and Goodaa Sunglasses, can provide users the ability to enjoy music as well as answer phone calls while keeping their hands free during walking, biking, or driving. Fitbit fitness watches can continuously monitor heart rate, track activity, and even provide on-screen workouts with no need to be manually configurated on a daily basis. With innovations on the horizon, the future of wearable devices will go beyond answering calls or counting our steps to providing us with sophisticated wearable gadgets capable of addressing fundamental and technological challenges.

This book investigates the development of wearable technologies across a range of applications from educational assessment to health, biomedical sensing, and energy harvesting. Furthermore, it discusses some key innovations in micro/nano fabrication of these technologies, their basic working mechanisms, and the challenges facing their progress.

Dr Noushin Nasiri
Head of NanoTech Laboratory,
School of Engineering,
Faculty of Science and Engineering,
Macquarie University,
Sydney NSW, Australia

Section 1

Health

Chapter 1

Introductory Chapter: Wearable Technologies for Healthcare Monitoring

Noushin Nasiri

1. Introduction

Wearable technologies are becoming increasingly popular as personal health system, enabling continuous real-time monitoring of human health on a daily basis and outside clinical environments [1–3]. The wearable device market is currently having a worldwide profit of around $34 billion and is expected to reach above $50 billion by 2022 owing to wearables' ease of use, flexibility, and convenience [4]. Real-time monitoring, operational efficiency, and fitness tracking are reported as main factors supporting the market growth of health wearable devices such as smart watches, smart glasses, and other wellness gadgets, with expected $12.1 billion world market by 2021 [5].

In the past decade, the recent progress in developing wearable devices was more focused on monitoring physical parameters, such as motion, respiration rate, etc. [3, 6, 7]. Today, there is a great interest in evolving wearable sensors capable of detecting chemical markers relevant to the status of health. Different approaches have been applied by researchers to design and fabricate wearable biosensors for remote monitoring of metabolites and electrolytes in body fluids including tear, sweat, and saliva [3, 8–10]. A great example would be the development of small and reliable sensors that would allow continuous glucose monitoring in diabetic patients [11, 12]. Diabetes is a chronic disease that can significantly impact on quality of life and reduce life expectancy. However, diabetics can stay one step ahead of the disease by monitoring their blood glucose level to minimize the complication of the disease by proper administration of insulin. Currently, blood analysis is the gold standard method for measuring the level of glucose in patient's blood. However, this technique cannot be applied without penetrating the skin, which can be painful and inconvenient, and requires user obedience. Therefore, current research focuses on the development of portable and wearable devices capable of continuous glucose sensing through noninvasive detection techniques.

2. Tear analysis

A majority of the recent studies in this field have targeted the area of personalized medicine, endeavoring to develop miniaturized wearable devices featuring real-time glucose monitoring in diabetic patients [12–15]. One great example is contact lens which is an ideal wearable device that can be worn for hours without any pain or discomfort [16]. Integration of glucose biosensors into contact lenses has recently been demonstrated by several research groups [9, 17, 18]. However, the level of glucose in tear fluid is very low (0.1–0.6 mM), requiring a high sensitivity of the sensor for picking up the signal from expected chemical reaction [3, 19]. Yao et al. [16] have fabricated a contact lens with integrated sensor for continuous tear glucose monitoring with wireless communication system over a distance of several centimeters. The sensor demonstrated a fast response of 20 s with a minimum detection of less than 0.01 mM glucose, which is 10–60 times lower than glucose level in human tear [16].

In addition to glucose, lactate is an important metabolite in the human body, which gets converted into L-lactate under hypoxic condition [20]. L-Lactate levels in tear fluid is about 1–5 mmol L^{-1}, which might increase significantly due to some heath conditions including ischemia, inadequate tissue oxygenation, stroke, and different types of cancer [21]. Thomas et al. [22] demonstrated an invasive detection of lactate in human tear by integrating an amperometric lactate sensor with Pt working (WE) and reference (RE) electrodes as well as a counter electrode (CE) as current drain, on a polymer-based contact lens, measuring lactate in situ in human tears without any need for physical sampling [22].

Very recently, Park et al. [17] reported a novel approach for fabricating fully transparent and stretchable smart contact lens capable of wirelessly monitoring the level of glucose in the tears of diabetic patients. **Figure 1** shows the layout of fabricated devices made of glucose sensors, wireless circuit, and display pixel on soft and transparent contact lens substrate (**Figure 1a** and **b**). The circuit diagram of the device is illustrated in Figure 1a, with radio frequency antenna receiving signals from a transmitter and a rectifier converting the signals to DC (**Figure 1a** and **c**). A continuous network of ultralong Ag nanofibers was used as stretchable electrodes for the antenna and interconnects (**Figure 1d**). In the case of any change in the concentration of glucose in tear, the sensor resistance changes resulting in the light-emitting diode (LED) pixel turning on or off. The device was tested in vitro using a live rabbit, providing substantial finding for smart contact lenses as one of the promising wearable devices in healthcare system [17].

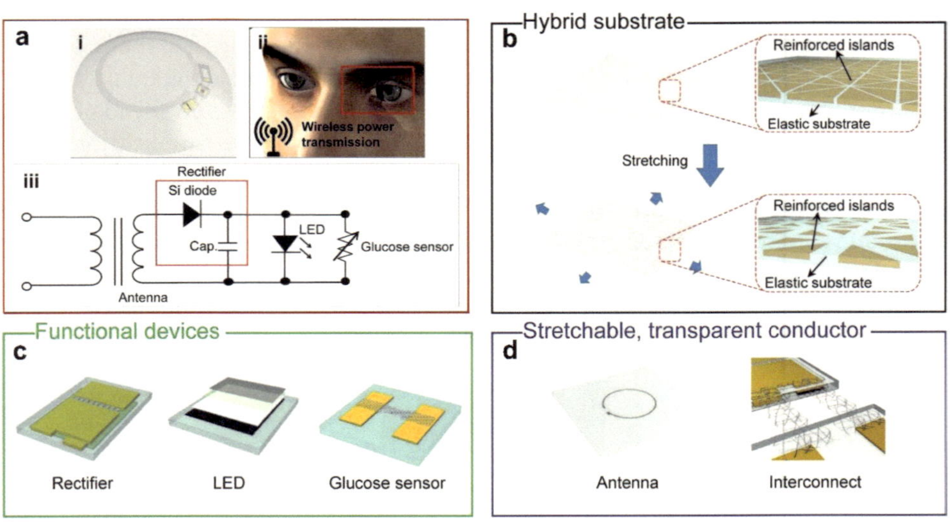

Figure 1.
(a) (i) Schematic illustration and (ii) operation of the soft, smart contact lens and (iii) the circuit diagram of the smart contact lens system. The soft, smart contact lens is composed of (b) a hybrid substrate; (c) functional devices including rectifier, LED, and glucose sensor; and (d) a transparent, stretchable conductor for antenna and interconnects [17].

3. Sweat analysis

In addition to tear, sweat electrolyte concentrations and blood serum are related [2, 8]. As one of the most readily accessible human biofluids, a great deal of information about the human body and its physical performance could be obtained via monitoring sweat electrolyte concentrations [23, 24]. Several groups have reported the key biomarkers in human sweat (e.g., sodium level, pH change, lactate concentration) relevant to human health and well-being, for monitoring athletic performance during sporting activities [25]. Jia et al. fabricated a skin-worn tattoo-based sensor for real-time monitoring of lactate in human sweat, offering substantial benefits for biomedical as well as sport applications [25]. In another approach, Curto et al. [26] fabricated a wearable and flexible microfluidic platform capable of monitoring changes in the sweat pH in real time. Anastasova et al. [27] developed a flexible microfluidic device for real-time monitoring of metabolite such as lactate as well as electrolytes such as pH and sodium in human sweat. Recently, Gao et al. [28] developed a flexible and wearable device (**Figure 2**) made of arrays of sensors for real-time monitoring of heavy metals, such as Zn, Cu, and Hg in human sweat. The device fabrication method is presented in **Figure 2a**, showing the deposition and stripping steps on microelectrodes. The sensing mechanism was based on an electrochemical detection of targeted heavy metals through four microelectrodes, including Au and Bi working electrodes, Ag reference electrode, and an Au counter electrode (**Figure 2b** and **c**). The fabricated device demonstrated high stability and selectivity toward heavy metals, providing a great platform to advancing the field of wearable biosensors for healthcare application, via monitoring the level of some heavy metals in human sweat [28]. A balanced level of Zn is necessary in the human body as a low and high Zn concentration can lead to pneumonia and liver damages, respectively [29, 30]. High level of Cu in the human body can lead to several diseases including Wilson's disease and heart, kidney, and liver failures as well as brain diseases [31, 32]. The fabricated device demonstrated high stability and selectivity toward heavy metals, providing a great platform to advancing the field of wearable biosensors for healthcare application [28].

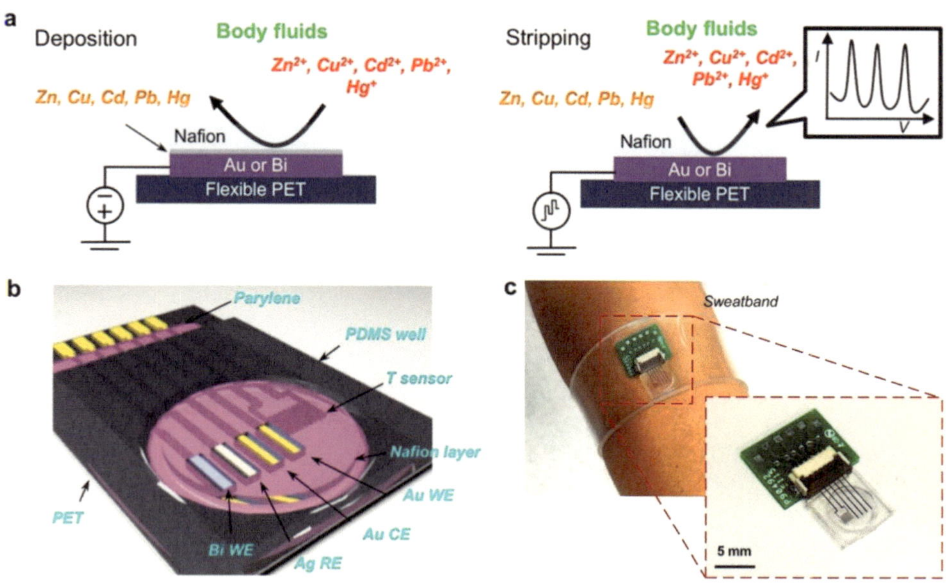

Figure 2.
(a) A schematic showing the concept of deposition and stripping on microelectrodes. (b) A schematic showing the composition of the microsensor array. (c) Optical image of a flexible sensor array interfacing with a flexible printed circuit connector [28].

4. Saliva analysis

Saliva, as a great diagnostic fluid, can be used in personal health devices for real-time monitoring of chemical markers including salivary lactate analysis [33]. Chai et al. developed a saliva nanosensor with a radio-frequency identification tag, integrated into dental implants for detecting cardiac biomarkers in saliva and predicting close heart attack in patients suffering from cardiovascular diseases [34]. In another approach, an instrumented mouthguard was designed and fabricated by Kim et al. [35] for measuring salivary uric acid levels which could be a biomarker for several diseases including hyperuricemia, gout, physical stress, and renal syndrome. The fabricated device showed high selectivity and sensitivity to low level of uric acid as well as great stability during a 4-h operation period [35]. Mannoor et al. [36] developed a hybrid biosensor made of graphene layers printed onto water-soluble silk, for noninvasive detection of bacteria through body fluids including sweat and saliva. This graphene/silk hybrid device illustrated an extremely high sensitivity to bacteria in body fluid with detection limits down to a single bacterium [36]. In addition, the fabricated device provided the potential users with battery-free operation and wireless communication system via radio frequency [36]. Arakawa et al. [37] designed and fabricated a salivary sensor equipped with a wireless measurement system, embedded onto a mouthguard support, featuring a high sensitivity toward detection of glucose over a range of 5–1000 μmol L^{-1}. The device demonstrated a great stability during a 5-h real-time glucose monitoring period in an artificial saliva with a phantom jaw [37]. In a similar approach, de Castro et al. [38] developed a microfluidic paper-based device integrated into a mouthguard, for continues monitoring of glucose and nitrite in human saliva. The saliva samples were collected from periodontitis and/or diabetes patients as well as healthy individuals. The fabricated device featured a low detection limit of 27 and 7 μmol L^{-1} for glucose and nitrite, respectively [38].

5. Summary

In summary, there is a great potential for micro- and nanosensors' integration into healthcare monitoring devices, developing new technologies for noninvasive detection of diseases in the human body. Flexible wearable devices offer promising capabilities in real-time monitoring of body fluids including tear, sweat, and saliva. However, more research is required to expand the use of wearable platforms in continuous analysis of body fluids, providing reliable real-time detection of targeting ions and proteins, among other complex analytes.

Author details

Noushin Nasiri
School of Engineering, Faculty of Science and Engineering, Macquarie University, Sydney, NSW, Australia

*Address all correspondence to: noushin.nasiri@mq.edu.au

References

[1] Trung TQ, Lee NE. Flexible and stretchable physical sensor integrated platforms for wearable human-activity monitoring and personal healthcare. Advanced Materials. 2016;**28**(22):4338-4372

[2] Nakata S, Arie T, Akita S, Takei K. Wearable, flexible, and multifunctional healthcare device with an ISFET chemical sensor for simultaneous sweat pH and skin temperature monitoring. ACS Sensors. 2017;**2**(3):443-448

[3] Tricoli A, Nasiri N, De S. Wearable and miniaturized sensor technologies for personalized and preventive medicine. Advanced Functional Materials. 2017;**27**(15):1605271

[4] Arnold JF, Sade RM. Wearable technologies in collegiate sports: The ethics of collecting biometric data from student-athletes. The American Journal of Bioethics. 2017;**17**(1):67-70

[5] Lymberis A, De Rossi DE. Wearable Ehealth Systems for Personalised Health Management: State of the Art and Future Challenges. Vol. 108. Netherlands: IOS Press; 2004

[6] Takei K, Honda W, Harada S, Arie T, Akita S. Toward flexible and wearable human-interactive health-monitoring devices. Advanced Healthcare Materials. 2015;**4**(4):487-500

[7] Honda W, Harada S, Arie T, Akita S, Takei K. Wearable, human-interactive, health-monitoring, wireless devices fabricated by macroscale printing techniques. Advanced Functional Materials. 2014;**24**(22):3299-3304

[8] Bariya M, Nyein HYY, Javey A. Wearable sweat sensors. Nature Electronics. 2018;**1**(3):160

[9] Kim J, Kim M, Lee M-S, Kim K, Ji S, Kim Y-T, et al. Wearable smart sensor systems integrated on soft contact lenses for wireless ocular diagnostics. Nature Communications. 2017;**8**:14997

[10] Malon RS, Sadir S, Balakrishnan M, Córcoles EP. Saliva-based biosensors: Noninvasive monitoring tool for clinical diagnostics. BioMed Research International. 2014;**2014**:962903

[11] Cappon G, Acciaroli G, Vettoretti M, Facchinetti A, Sparacino G. Wearable continuous glucose monitoring sensors: A revolution in diabetes treatment. Electronics. 2017;**6**(3):65

[12] Kim J, Campbell AS, Wang J. Wearable non-invasive epidermal glucose sensors: A review. Talanta. 2018;**177**:163-170

[13] Arakawa T, Kuroki Y, Nitta H, Toma K, Mitsubayashi K, Takeuchi S, et al. Mouth guard type biosensor "cavitous sensor" for monitoring of saliva glucose with telemetry system. In: 2015 9th International Conference on Sensing Technology (ICST), 2015. New Zealand: IEEE; 2015. pp. 46-49

[14] Blaikie TP, Edge JA, Hancock G, Lunn D, Megson C, Peverall R, et al. Comparison of breath gases, including acetone, with blood glucose and blood ketones in children and adolescents with type 1 diabetes. Journal of Breath Research. 2014;**8**(4):046010

[15] Chu MX, Miyajima K, Takahashi D, Arakawa T, Sano K, Sawada S-I, et al. Soft contact lens biosensor for in situ monitoring of tear glucose as non-invasive blood sugar assessment. Talanta. 2011;**83**(3):960-965

[16] Yao H, Liao Y, Lingley A, Afanasiev A, Lähdesmäki I, Otis B, et al. A contact lens with integrated telecommunication circuit and sensors for wireless and continuous tear glucose monitoring. Journal of

Micromechanics and Microengineering. 2012;**22**(7):075007

[17] Park J, Kim J, Kim S-Y, Cheong WH, Jang J, Park Y-G, et al. Soft, smart contact lenses with integrations of wireless circuits, glucose sensors, and displays. Science Advances. 2018;**4**(1):eaap9841

[18] Badugu R, Lakowicz JR, Geddes CD. Noninvasive continuous monitoring of physiological glucose using a monosaccharide-sensing contact lens. Analytical Chemistry. 2004;**76**(3):610-618

[19] Lin Y-R, Hung C-C, Chiu H-Y, Chang P-H, Li B-R, Cheng S-J, et al. Noninvasive glucose monitoring with a contact lens and smartphone. Sensors. 2018;**18**(10):3208

[20] Farandos NM, Yetisen AK, Monteiro MJ, Lowe CR, Yun SH. Contact lens sensors in ocular diagnostics. Advanced Healthcare Materials. 2015;**4**(6):792-810

[21] Pankratov D, González-Arribas E, Blum Z, Shleev S. Tear based bioelectronics. Electroanalysis. 2016;**28**(6):1250-1266

[22] Thomas N, Lähdesmäki I, Parviz BA. A contact lens with an integrated lactate sensor. Sensors & Actuators, B: Chemical. 2012;**162**(1):128-134

[23] Liu G, Ho C, Slappey N, Zhou Z, Snelgrove SE, Brown M, et al. A wearable conductivity sensor for wireless real-time sweat monitoring. Sensors & Actuators, B: Chemical. 2016;**227**:35-42

[24] Pribil MM, Laptev GU, Karyakina EE, Karyakin AA. Noninvasive hypoxia monitor based on gene-free engineering of lactate oxidase for analysis of undiluted sweat. Analytical Chemistry. 2014;**86**(11):5215-5219

[25] Jia W, Bandodkar AJ, Valdés-Ramírez G, Windmiller JR, Yang Z, Ramírez J, et al. Electrochemical tattoo biosensors for real-time noninvasive lactate monitoring in human perspiration. Analytical Chemistry. 2013;**85**(14):6553-6560

[26] Curto VF, Fay C, Coyle S, Byrne R, O'Toole C, Barry C, et al. Real-time sweat pH monitoring based on a wearable chemical barcode micro-fluidic platform incorporating ionic liquids. Sensors & Actuators, B: Chemical. 2012;**171-172**:1327-1334

[27] Anastasova S, Crewther B, Bembnowicz P, Curto V, Ip HM, Rosa B, et al. A wearable multisensing patch for continuous sweat monitoring. Biosensors & Bioelectronics. 2017;**93**:139-145

[28] Gao W, Nyein HY, Shahpar Z, Fahad HM, Chen K, Emaminejad S, et al. Wearable microsensor array for multiplexed heavy metal monitoring of body fluids. ACS Sensors. 2016;**1**(7):866-874

[29] Lassi ZS, Moin A, Bhutta ZA. Zinc supplementation for the prevention of pneumonia in children aged 2 months to 59 months. Cochrane Database of Systematic Reviews. 2016;**12**:CD005978

[30] Mohammad MK, Zhou Z, Cave M, Barve A, McClain CJ. Zinc and liver disease. Nutrition in Clinical Practice. 2012;**27**(1):8-20

[31] Crisponi G, Nurchi VM, Fanni D, Gerosa C, Nemolato S, Faa G. Copper-related diseases: From chemistry to molecular pathology. Coordination Chemistry Reviews. 2010;**254**(7-8):876-889

[32] Huster D. Wilson disease. Best Practice & Research. Clinical Gastroenterology. 2010;**24**(5):531-539

[33] Lee J, Garon E, Wong D. Salivary diagnostics. Orthodontics & Craniofacial Research. 2009;**12**(3):206-211

[34] Chai PR, Castillo-Mancilla J, Buffkin E, Darling C, Rosen RK, Horvath KJ, et al. Utilizing an ingestible biosensor to assess real-time medication adherence. Journal of Medical Toxicology. 2015;**11**(4):439-444

[35] Kim J, Imani S, de Araujo WR, Warchall J, Valdés-Ramírez G, Paixão TRLC, et al. Wearable salivary uric acid mouthguard biosensor with integrated wireless electronics. Biosensors & Bioelectronics. 2015;**74**:1061-1068

[36] Mannoor MS, Tao H, Clayton JD, Sengupta A, Kaplan DL, Naik RR, et al. Graphene-based wireless bacteria detection on tooth enamel. Nature Communications. 2012;**3**:763

[37] Arakawa T, Kuroki Y, Nitta H, Chouhan P, Toma K, Sawada S-I, et al. Mouthguard biosensor with telemetry system for monitoring of saliva glucose: A novel cavitas sensor. Biosensors and Bioelectronics. 2016;**84**:106-111

[38] de Castro LF, de Freitas SV, Duarte LC, de Souza JAC, Paixão TR, Coltro WK. Salivary diagnostics on paper microfluidic devices and their use as wearable sensors for glucose monitoring. Analytical and Bioanalytical Chemistry. 2019:1-10

Chapter 2

Noninvasive Acquisition of the Aortic Blood Pressure Waveform

Mart Min, Hip Kõiv, Eiko Priidel, Ksenija Pesti and Paul Annus

Abstract

Blood pressure reflects the status of our cardiovascular system. For the measurement of blood pressure, we typically use brachial devices on the upper arm, and much less often, the radial devices with pressure sensors on the wrist. Medical doctors know that this is an unfortunate case. The brachial pressure and even more, the radial pressure, both are poor replacements for the central aortic pressure (CAP). Moreover, the devices on the market cannot provide continuous measurements 24 h. In addition, most of the ambulatory and wearable monitors do not enable acquisition of the blood pressure curves in time. These circumstances limit the accuracy of diagnosing. The aim of this chapter is to introduce our experiments, experiences and results in developing the wearable monitor for central aortic blood pressure curve by using electrical bioimpedance sensing and measurement. First, electronic circuitry with embedded data acquisition and signal processing approaches is given. Second, finding appropriate materials, configurations and placements of electrodes is of interest. Third, the results of modelling and simulations are discussed for obtaining the best sensitivity and stability of the measurement procedures. Finally, the discussion on the provided provisional experiments evaluates the obtained results. The conclusions are drawn together with the need for further development.

Keywords: blood pressure waveform, central aortic pressure, cardiovascular system, medical indications, diagnosing, electrical impedance, bioimpedance-based sensing, modelling, simulation, signal processing, transfer function, noninvasive measurements, electrodes, wearable devices

1. Introduction

Hypertension, one of the most common medical disorders, "silent killer" and tremendous global burden, is often overlooked until otherwise healthy individual has a regular health check and the doctor discovers that the blood pressure (BP) is too high. What does it mean? Typically, it means that the systolic blood pressure (SBP) is over 140 mmHg and diastolic blood pressure (DBP) is over 90 mmHg. This probably implies that the heart is under huge load and it can soon wear out causing cardiovascular diseases (CVD). Systolic pressure indicates how much pressure the blood is exerting against artery walls when the heart beats. Diastolic blood pressure shows how much pressure is exerted while the heart is in a resting condition between the beats. As large arteries start to stiffen with age, it is normal that SBP rises; but for every 20 mmHg systolic or 10 mmHg diastolic increase in BP, there is a doubling of mortality from ischemic heart disease and stroke [1].

1.1 Why an aortic pressure?

Brachial cuff sphygmomanometer is widely used to assess the pressure parameters for both, diagnosis and treatment decisions. This is unfortunate, as we have known for over a half a century that brachial pressure is a poor surrogate for central aortic pressure (CAP), which is invariably lower than corresponding brachial values [2]. It is quite logical that CAP represents the true load imposed on the heart and large arteries rather than the brachial [3], but there are many reasons why we are still so stuck in old methods. For example, lack of proper guidelines for alternative technologies (noninvasive and nonocclusive) which are not standardised as brachial pressure assessment with oscillometric devices are. This makes it hard to trust the new ones. In addition, the cuff method is easy and quick for the doctors to execute. There is still a lot of work to do to prove that the cardiovascular (CV) risk stratification and monitoring response to therapy are better when based on central rather than brachial pressure [2].

Profound study on accuracy of cuff-measured blood pressure was conducted in 2017, and they found that, on average, cuff BP underestimates intraarterial brachial systolic BP by 5.7 mmHg and overestimates diastolic systolic BP by 5.5 mmHg [4, 5]. This means that the real load on the heart is often unknown and wrongly evaluated causing questionable treating decisions. In addition, the systolic and diastolic pressures shown only as two numbers do not reflect the situation of the patient as well as a real-time and continuous aortic pressure waveform. Continuous aortic waveform provides important information about derived parameters, which are intrinsically created by the pulse pressure profile, like left ventricular stroke volume, cardiac output, vascular resistance and pulse pressure variations in real time [6].

If we are talking about different devices and methods to assess BP, the invasive intraarterial catheterization ("gold standard") is undoubtedly the most accurate and reliable technique to consider, providing continuous and beat-by-beat reading of blood pressure variations. During cardiac catheterization to diagnose CV condition and at the same time to measure the pressure directly in the aorta, a thin tube is inserted into an artery (radial, femoral or brachial) and threaded to the heart. Albeit major complications are uncommon, the procedure can cause infections, nephropathy, cholesterol emboli, local vascular injury, hematoma and arteriovenous fistula [7]. Unfortunately, this procedure is not feasible for daily use, as it is technically demanding and costly, requiring well-trained personnel. Preventing hypertension would be much more efficient if we could observe the BP during people's everyday life. When a doctor has a suspicion that the patient has hypertension, they do the ambulatory BP monitoring. The patient gets the brachial sphygmomanometry device for 24-h home monitoring to measure the blood pressure by inflating the cuff multiple times per day and during the night. In addition to the fact that cuff measurement variably under- or overestimates SBP at the aorta [5], the measurement is very uncomfortable for long-time recording due to pressing the blood flow shut.

1.2 What are the alternatives?

Karamanoglu et al. presented in 1993 that it is possible to use transfer function between the ascending aorta and the brachial or radial artery to estimate central (aortic) blood pressure (CAP) [8]. This work opened the door for noninvasive method called applanation tonometry. Currently it has the widest application in devices that perform pulse wave analysis and assessment of aortic pressure waveform [9]. Tonometry sensor probe flattens the artery so that transmural forces within the vessel wall are perpendicular to the arterial surface to measure pressure transmitted through the skin [10]. Preferred measuring site is often the radial artery, as the measurements are more easily reached when there is a firm base under the soft artery (radius bone under radial artery). One of the earliest arterial tonometry apparatus on the market that is clinically approved by the US Food and Drug Administration (FDA) is SphygmoCor by AtCor Medical, Sydney, Australia [11]. SphygmoCor was also the first device that used the general transfer function to estimate CAP waveform from peripheral arteries [12]. It uses pencil-like sensor that is pressed against the artery by trained healthcare professional, but a number of limitations occur. Firstly, due to manual positioning of the tonometer over the artery, the readings can be operator dependent. Secondly, it can be difficult to obtain high-quality pulse in some subjects with lower blood pressure or with obesity. Thirdly, tonometry requires calibration with brachial cuff technique, and finally, the blood vessel is flattened against the bone and possibly disturbing the blood flow [11, 13].

There are not many reliable ambulatory devices that could measure the aortic pressure or aortic pressure waveform in 24 h. Operator-independence would be a tremendous step forward in assessing patients with prehypertension, as evaluating patient's BP change during the day and night gives a good insight into the scope of the disease. BPro (HealthSTATS International, Singapore) is a wristwatch-type BP sensor that has come strongly into the market lately. It acquires radial pressure waveform through automated radial tonometry, and the software estimates aortic pressure values using N-point moving average method, but it does not provide an aortic waveform [14]. This wearable device is not fully accepted in regular clinical work as the accuracy and reliability are still unclear. Study by Harju et al. discovered inaccurate readings [15] when comparing the BPro device with standard invasive monitoring, based on recommendations by the Association for the Advancement of Medical Instrumentation (AAMI). On a Bland-Altman plot, the bias and precision between these two methods was 19.8 ± 16.7 mmHg [15]. So there is a room for development, especially in the world of wearables that measure blood pressure. Even though there are numerous different methods established already, it is not an easy task to tackle.

2. Bioimpedance-based sensing

Another, less known technique to derive a blood pressure-related waveform from the radial artery is bioimpedance. Small current is applied to the interested site through electrodes, and a voltage difference is measured (**Figure 1**). Bioimpedance is calculated from the exerted current and measured voltage, which gives us the change of the impedance during cardiac cycle. With each heartbeat, the volume of the blood changes under the electrodes, and it reflects in impedance curve which corresponds to blood pressure waveform. The measurement site is often the radial artery, because the signal source is closest to the skin. Our workgroup has discovered that similarly with tonometer, applying the developed general transfer function to the measured signal, we can assess central aortic blood pressure at least as well as with tonometry device.

Bioimpedance sensing does not need strong pressure on the artery, as it is with tonometry, only a permanent electrical contact is required. Therefore, the worry of affecting the blood circulation with the measurement procedure falls off in a large extent—the measurements become more passive. The first papers that suggested the viability of bioimpedance measurements for pressure assessment circulated already in the 1980s. Herscovici and Roller [16] proposed in 1986 a possibility to determine the mean arterial pressure with impedance plethysmography by attaching four conductive Velcro electrodes to the regular blood pressure cuff. The algorithm

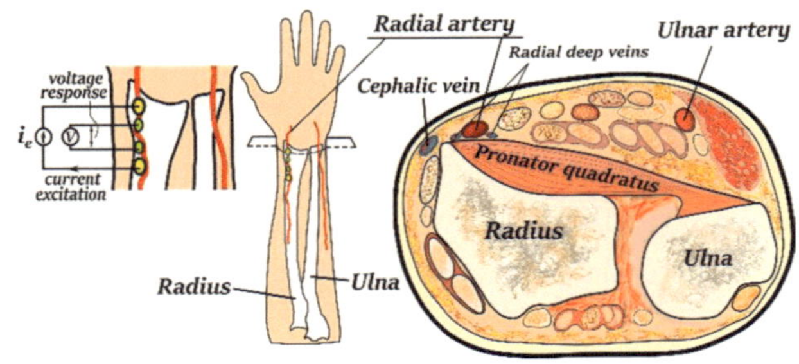

Figure 1.
Four electrodes placed on the radial artery and the cross-section of the wrist.

applied to find the central pressure value showed a good correlation between direct measurement of intra-aortic pressure curve and indirect impedance signal. In 1994, Rudolf A. Hatschek [17] patented a blood pressure measuring device and method, which allows to make measurements in a noninvasive manner. He explains that the blood pressure can be determined relatively accurately by obtaining two different values: blood volume, as a variable that changes periodically over time in the rhythm of the pulse beat, and a pulse wave velocity. By linking these two values together, it is possible to form at least one blood pressure value or its change (systolic pressure, diastolic pressure or the average blood pressure). Among other proposed possibilities as light waves, ultrasonic waves and magnetic/electrical induction, Hatschek suggests to configure the device so that it determines the changing blood volume in the measuring region of a body part with the electrical impedance. Japanese workgroup's patent application [18] was published in 2010 for a device that measures the pulse wave of a radial artery and among other parameters as cardiac load and hardness of artery, also a blood pressure value derived from the pulse wave of the artery. The device consists of four electrodes placed on a cuff, and it detects the blood volume fluctuation of the radial artery as the variations in electrical bioimpedance (EBI) to acquire the volume pulse wave. Solà et al. presented in 2011 a pilot study [19], where they provided first experimental evidence that electrical impedance tomography (EIT) is capable of measuring pressure pulses directly within the descending aorta. Their research measures the impedance on the thorax, not on the arm or wrist, but the study supports, nevertheless, the idea of central aortic pressure assessment with bioimpedance. Recently, He et al. published a promising paper [20] in 2016, which discusses pulse wave detection method based on the bioimpedance of the arteria radialis. The aim of this paper is to analyse the impedance pulse wave to obtain the pulse rate, but refers also to the central aortic pressure waveform. A number of researchers have had analogs thoughts and promising results, and a number of scholars have had practical results in improvement of the EBI-based measurements of aortic pressure curve. At the same time, the development of corresponding devices for clinical practice is still not significant. Nevertheless, the interest to get a blood pressure measurement device that relies on bioimpedance is still very topical. Especially, when the big corporate, Microsoft Technologies, got their patent published in 2018 for a wearable system that determines a pulse waveform based on bioimpedance measurement device together with pressure transducer [21].

2.1 Bioimpedance measurement device

For the measurement of bioimpedance variations (bio-modulation) at the wrist on top of the radial artery, a wearable device was designed. The work principle consists of generating a single-frequency sinewave from an excitation current source through the impedance and detecting the voltage response to it synchronously with excitation current (lock-in demodulation). A 12-bit digital-to-analog converter (DAC) generates the constant value excitation current in the frequency range from 1 to 100 kHz from a digital waveform, and a differential input instrumentation amplifier picks up the voltage response from the impedance. A two-phase (0 and 90°) synchronous rectifier demodulates and separates the response voltage V_{RES} into real (Re) and quadrature (Im) components. Two 32-bit analog-to-digital converters (ADC) digitise both the components for further signal processing and communication. The simplified block diagram of the device is given in **Figure 2**. **Figures 3** and **4** show a photo of the prototyped solution.

Experimental circuitry uses the state-of-the-art linear technology/analog devices LTC2508-32 32-bit over-sampling ADC, which is reasonably low-noise and low-power micro device, containing embedded configurable filter for digital averaging and noise smoothing. Direct conversion of the impedance signal is not possible anymore, since the high-resolution 32-bit ADC's are relatively slow. Classical synchronous demodulators were introduced in the path, and only the slowly varying bio-modulation $\Delta Z(t)$ was left for the ADC instead of the high-frequency measurement signal.

The device was designed to have very low energy consumption, small footprint and good connectivity, all essential parameters for the wearable use. Bluetooth Low Energy (BLE) Version 4.0+ was used for the connectivity with host devices, and the power was supplied from the lithium-ion (Li-ion) battery. USB connection switches on only during charging the internal energy source. AVR microcontroller ATXMEGA256A3U with low energy consumption handled all the computing and communication tasks on the module.

2.2 Design consideration for a measurement system

The value of the measured bioimpedance Z is varying, but the base value of it, Z_0, is huge compared to the information carrying modulation $\Delta Z(t)$ (see Eq. (1)

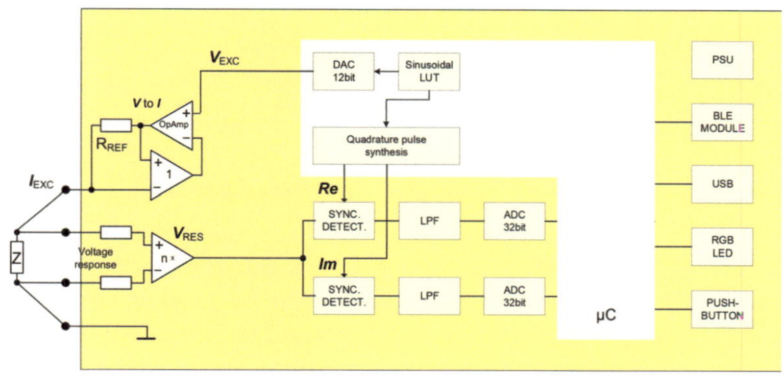

Figure 2.
Block diagram of the measurement device.

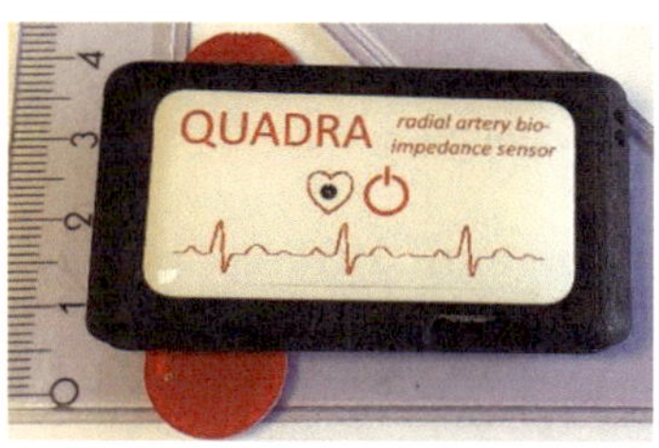

Figure 3.
The prototyped version of a wearable impedance measurement device.

Figure 4.
Image of the 32-bit impedance acquisition system prototype.

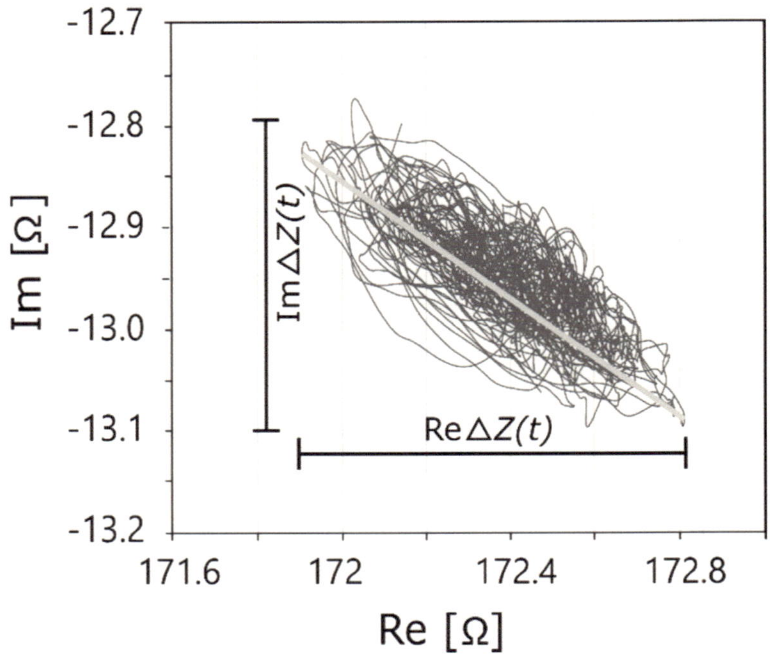

Figure 5.
Measured bioimpedance values on top of the radial artery on the Nyquist diagram [23].

in Chapter 3.1). Several observations can be drawn from the generalized measurement results in **Figure 5**. The first observation is that the greyish, slightly smeared information carrying signal, ΔZ(t), is tiny compared to the base value Z_0 of the acquired bioimpedance. Next, the imaginary part Im Z of the impedance vector Z is nearly 10 times smaller than the real part Re Z. Third, the modulation is roughly in the direction of the vector of the impedance base value Z_0. The conclusion is that the role of Im Z is low, and less attention can be paid into the accuracy of the vector measurements when designing the device, especially the synchronous detector of it. A root problem in designing a suitable electronics is whether to use analog or digital realisation of the synchronous detector.

2.3 A novel solution measurement of differences

Certified medical bioimpedance measurement device CircMon BT101, which we have been using so far during the clinical studies, employs a carrier (base value Z_0) compensation method [22]. The biggest drawback of this solution is the complexity of both, electronic circuitry and algorithms, for adjusting the compensation signal.

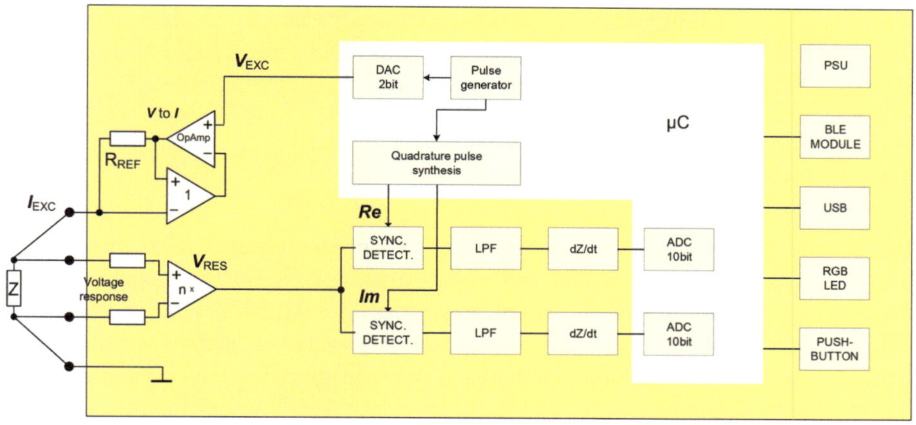

Figure 6.
Block diagram of the derivative bioimpedance signal acquisition system.

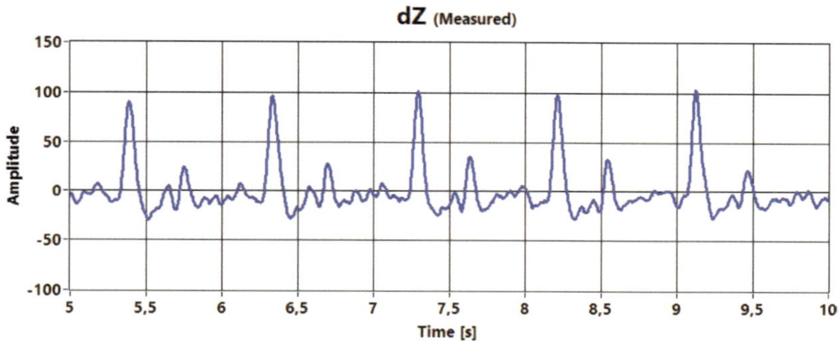

Figure 7.
The differenced bioimpedance signal dZ(t) measured by 10-bit ADC in numberic value range from −128 to +256. The measured waveform is smoothed by the third-order Savitzky-Golay filter within 10 sidepoints.

Increased energy consumption is a penalty for improvements. The conclusion is that a new, more effective solution should be developed.

The difference method has been introduced recently for achieving the same result [22, 23]. The main idea behind it lies in digitising of the difference between two consecutive samples instead of direct digitalisation of all the samples. It corresponds to taking a derivative, mathematically. Since the derivative of the constant is zero, the base value Z_0 of the impedance Z is eliminated, but the informative variations $\Delta Z(t)$ are upraised.

The novel test device in **Figure 6** contains AVR ATXMEGA microcontroller together with BLE 4.0+ module, but instead of high-quality external 32-bit ADC, an internal rather noisy low-quality 10-bit-embedded ADC was used.

The acquired signal presents now a derivative of the original biological impedance depicted in **Figure 7**. The original signal is restored by digital integration (**Figure 17**), which brings in an additional smoothing effect improving the resulting signal-to-noise ratio (SNR). At the same time, this integration may well be unnecessary in some cases.

Occasionally, one of the signal processing steps after acquiring the impedance signal is differentiation for finding certain peculiar points in the waveform. The first, second and even higher derivatives are used to find and calculate relevant cardiovascular parameters. In that sense, the acquired impedance signal in **Figure 7** is well suited without the integration step. The CAP waveform can also be derived directly from the derivative of the $\Delta Z(t)$. The experiments showed the presence of significant noise and disturbance when making provisional experiments with simple stainless steel electrodes [24], and the role of movement artefacts was highly troubling. This implies that the electrode design must be considered more seriously in further research.

3. Electrodes

Electrodes play a crucial role in bioimpedance measurements. The sensitivity to tiny impedance changes, as well as stability and repeatability of measurements depend on the quality of the electrodes. Bioimpedance variability during cardiac cycle is usually measured with two pairs of electrodes: two current-injecting electrodes and two voltage-sensing electrodes. This configuration cancels out electrode polarisation impedances and reduces dramatically the influence of skin-electrode contact resistance. However, quite frequently we cannot see this advantage, and the electrode-skin contact impedance remains prominent exceeding the actual bioimpedance of interest, which greatly affects the end signal quality [25]. This is especially important when measuring heartbeat-associated impedance variations from the wrist area, where they are minuscule (order of mΩ). Choosing appropriate electrodes increases the correct result probability, but the top skin layer (*stratum corneum*) against the electrode is very dry and badly conductive making electrode design extremely complicated. The main type to consider for bioimpedance measurements is disposable non-polarizable and pre-gelled silver/silver chloride (Ag/AgCl) electrodes. Pre-gelled electrodes have usually the lowest skin-electrode impedance, low motion artefacts and low noise level [26]. Unfortunately, as they are suitable for single use only, we do not consider them for wearable devices. Dry electrodes are a more prospective choice, but due to lack of gel between the skin and the electrode, there exists a significant capacitive layer, which increases the total impedance and the probability of motion artefacts [27].

3.1 Electrode placements and materials

The total impedance Z of the wrist consists of the invariable basal impedance Z_0 and a variable part $\Delta Z(t)$ that is caused by the pulse wave. As a result, the impedance expresses as

$$Z(t) = Z_0 + \Delta Z(t) \tag{1}$$

In order to detect the cardiac activity, the interesting variable is the $\Delta Z(t)$, assumedly reflecting the volume change of pulsating blood in arteries. A custom-made flexible electrode (**Figure 8a** and **b**) was used, positioned distally (**Figure 8c**) and circularly (**Figure 8d**) on top of the location of the radial artery.

Suitable materials for electrodes must be found and thoroughly tested for truly unobtrusive and reliable pervasive monitoring. Easy applicability is paramount. They should not irritate the skin; their parameters should stay reasonably unchanged during the acquisition cycle and should be insensitive to motion-induced stress.

In order to evaluate the effect of distal and circular placement of electrodes on the radial artery to the measured values of Z and $\Delta Z(t)$, the experiments were carried out having the excitation signal with the amplitude of 500 mV in the frequency range of 10–5000 kHz. The results are visible on **Figure 9**. The $\Delta Z(t)$ is few times higher in the case of longitudinal placement of electrodes (**Figure 9a**, red line) than in the case of transverse placement (**Figure 9a**, blue line). The total impedance Z is on average about 2.7 Ω higher in the case of transverse placement than in the case of longitudinal placement. When Z is decreasing with frequency, the $\Delta Z(t)$ is maintaining its relative

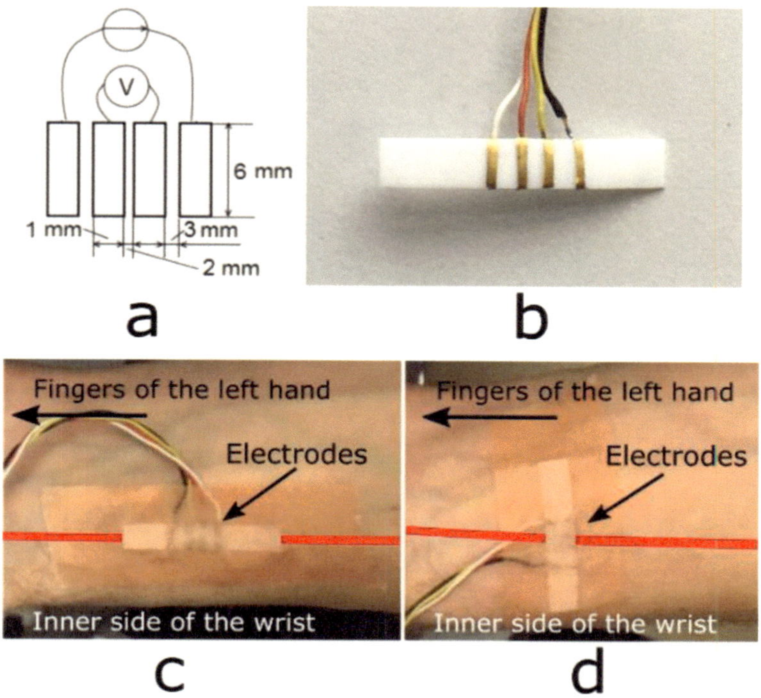

Figure 8.
Dimensions (a), design (b), and placement of a custom-made flexible four-electrode system in the case of a distal (c) and circular (d) locations on the wrist, where the thick red line denotes the approximate location of the radial artery (reprinted from [27]).

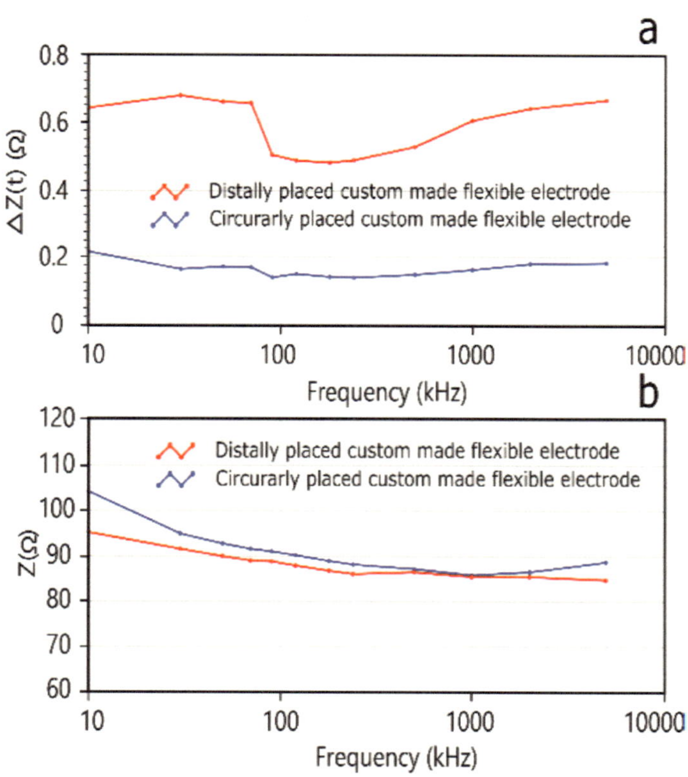

Figure 9.
Frequency response of measured (a) $\Delta Z(t)$ and (b) Z of the wrist in the cases of distal and circular placements of the electrodes.

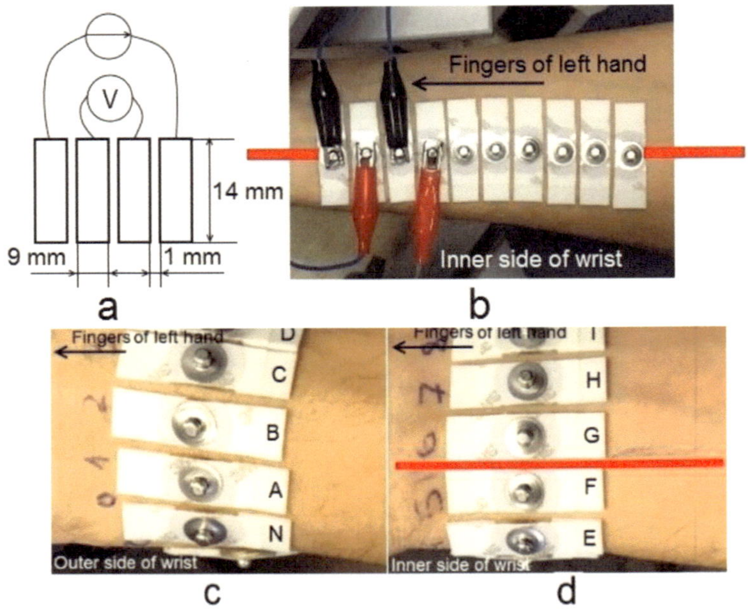

Figure 10.
Modified dimensions of the standard ECG electrodes in utilised four-electrode system (a) and the placement on the wrist in distal (b) and circular configuration (c and d), where the thick red line denotes the approximate location of the radial artery (reprinted from [27]).

value regardless of the excitation frequency. We can say that the longitudinal placement of electrodes possesses better results concerning the monitoring of cardiac activity in the wrist by using the prepared flexible electrode [28].

In order to verify the results of the custom-made electrode system, a similar research was performed by using the standard Ag/AgCl electrodes with foam tape (Type 2228 of 3M Health Care). Electrode dimensions were reduced physically (**Figure 10a**) and placed on the wrist distally (**Figure 10b**) and circularly (**Figure 10c** and **d**). The results in the case of distal placement of Ag/AgCl gel electrodes confirm the outcome of the results obtained with the custom-made electrodes.

Another custom-made electrode material was tested to try to improve the signal acquisition. Highly conductive carbon-based fillers added to the soft and flexible polydimethylsiloxane (PDMS) or silicone rubber matrix make a prospective dry electrode material. These fillers can be carbon nanotubes (CNTs), carbon nanofibres (CNFs), carbon fibres (CFs) and carbon black (CB). Previous researches have shown that these composites are biocompatible, and the existence of sweat and long-term wearing has little influence on the performance [27, 29]. We have developed a CNF/CF-PDMS material that could be used as electrodes for our wearable bioimpedance device due to its softness and stretchability [30]. Stratum corneum has very high impedance due to a large number of dead skin cells. Our hypothesis is that the developed electrode material can overcome this problem because the long fibres of carbon inside the silicone are sticking out and pressing a little bit into the skin layer (**Figure 11**).

We compared three different sets of electrodes: (a) Ag/AgCl gel electrodes, (b) carbon nanofibre electrodes (CNF-PDMS) and (c) carbon nanofibre together with carbon fibre electrodes (CNF/CF-PDMS). We abraded the skin slightly with a rough cloth for a better contact and placed the material on the wrist to register impedance variability with MFLI Lock-In amplifier (Zürich Instruments) on the frequency of 1o kHz. The results are shown on **Figure 12**. Pre-gelled commercially available electrodes showed good clean signal with impedance change of 0.1%. As skin-electrode

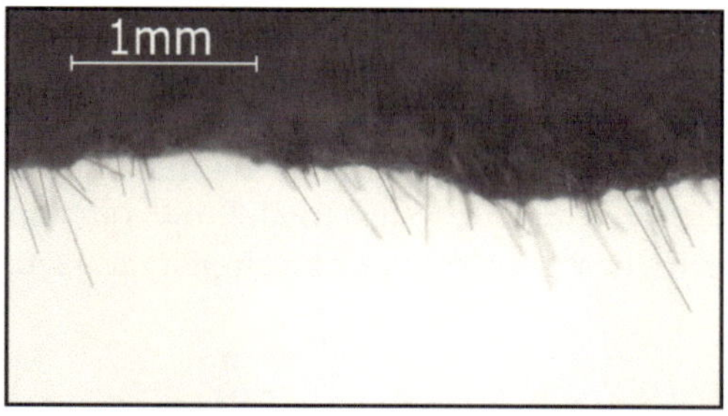

Figure 11.
Carbon fibre strands sticking out of the base material (CNF/CF-PDMS polymer) (reprinted from [30].

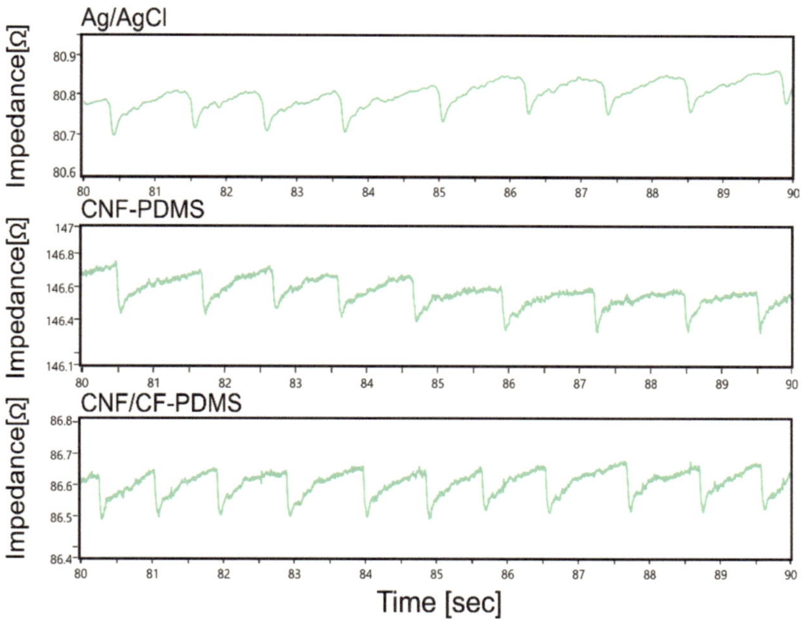

Figure 12.
Three different impedance signals from the wrist with Ag/AgCl, CNF-PDMS and CNF/CF-PDMS electrodes.

contact is worse, the CNF-PDMS and CNF/CF-PDMS soft electrodes gave slightly noisier signal, but the impedance change is clearly visible.

During these preliminary experiments, we could conclude that the CNF/CF-PDMS electrode material gave more stable results than CNF-PDMS over longer period of time. Further work needs to be done to establish whether silicone polymer together with carbon fibre and carbon nanofibre has a prospect to be used as electrodes for bioimpedance wearable devices. Also the question of the source of the signal arise—in what amount the blood itself contributes to the measured $\Delta Z(t)$ and in what amount it is caused by the rhythmical compression of tissues nearby [31].

4. Simulations

The development of mathematical and physical models of a haemodynamics is of great importance for the cardiovascular research [32]. The model is a simplified approximation of the real system, which incorporates most of the features. By using simulations, it is possible to predict the performance of the instrumentation, optimise and minimise the design and cost.

Noninvasive sensing instruments for bioimpedance measurement on the radial artery are highly sensitive to noise, and small errors on the measured data could turn into large mistakes in the final results [33]. In order to optimise the signal acquisition of the approach and to understand the impact of arterial pulse propagation to the results, it is reasonable to use modelling and simulation. In addition, we can determine the highest sensitivity of bioimpedance sensing on the radial artery.

4.1 Simulation of sensitivity distribution for EBI measurement

Sensitivity field is a frequently discussed topic in the impedance measurements. The transfer impedance (Z) can be approximated as the ration measured between the pick-up (PU) couple voltage (E) and the injected current (I) between the current-carrying (CC) couple [34].

$$Z = \frac{E}{I} \tag{2}$$

As biological tissue is inhomogeneous, the total measured impedance (Z) is the sum of all local resistivity (ρ) values of all small sub-volumes in the sample and can be written as following [35]:

$$Z = \iiint \rho \cdot \frac{J_{CC} \cdot J_{PU}}{I_{CC} \cdot I_{PU}} d \tag{3}$$

The sensitivity (S) of an impedance measurement is the scalar value representing the CC current density lines J_{CC} projection on the PU current density lines J_{PU} [35].

$$S = \frac{J_{CC} \cdot J_{PU}}{I_{CC} \cdot I_{PU}} \tag{4}$$

S is a positive value if measured impedance Z increases and negative if measured impedance Z decreases [35]. The sensitivity field S can be expressed by the following equation [36]:

$$S = J'_{reic} \cdot J'_{cc} \tag{5}$$

where J'_{cc} is current density and $J'_{reci} J'_{\mathrm{reci}}$ reciprocal density.

The sensitivity field in EBI measurements depends on several parameters like electrode number and geometry, orientation, configuration and spacing between electrode couples. Several configuration strategies have been published and researched for EIT applications. Some of these are neighbouring method [37], cross method [38], opposite method [33], adaptive method [39] and focused impedance measurement (FIM) [40].

4.2 Experimental bioimpedance sensitivity simulation

A finite element modelling (FEM) was used for simulation of four-electrode impedance measuring on the human forearm with different setups and configurations between electrode couples. The objective of the study was to describe the spatial sensitivity field in order to optimise the bioimpedance measurement acquisition of haemodynamics [34]. Two most common approaches of electrode placement for EBI measurements on the wrist are distal and circular [28] which are also used for simulation. The sensitivity can be represented as a projection of the density lines of current-carrying electrode couple on the voltage pick-up density lines,

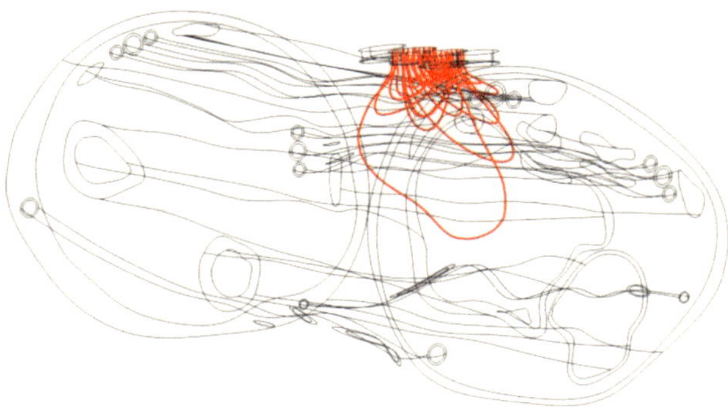

Figure 13.
Simulated current density lines on the wrist (reprinted from [33]).

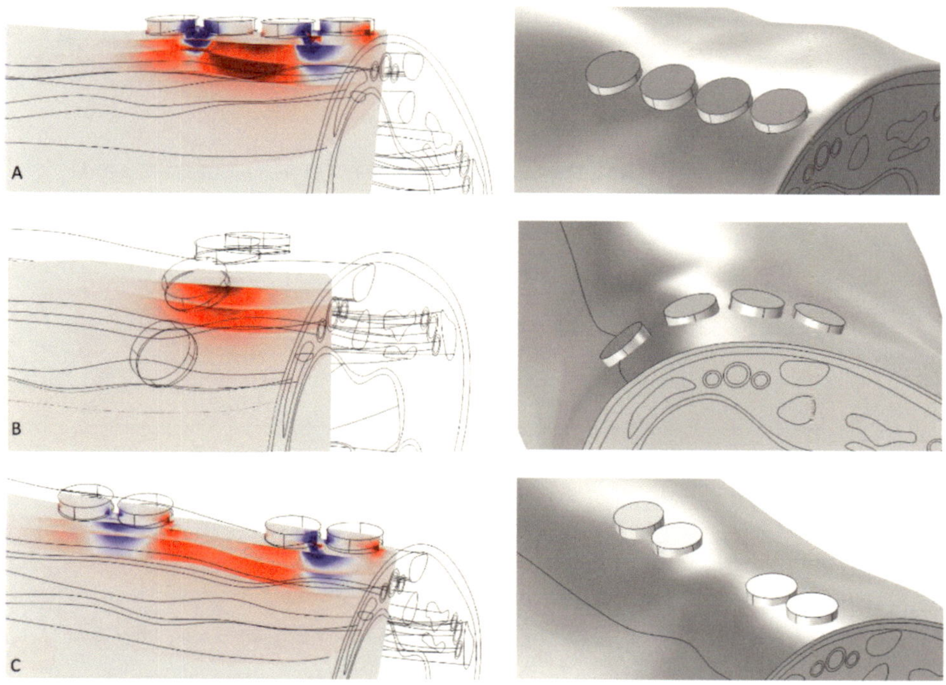

Figure 14.
Examples of calculated sensitivity maps obtained for different electrode configurations. Four electrodes are placed on the wrist in distal (a and c) and circular (b) configuration. Scaling of the colour map is kept the same within each simulation. The sensitivity is shown in the colour map: Positive values are indicated with red colour, and negative values are indicated with blue colour (reprinted from [33]).

and it describes how effectively different regions are contributing to the measured signal (**Figure 13**) [33]. In **Figure 14**, the configurations A and C have regions of both positive and negative sensitivities, but B (electrodes circularly) detects that the radial artery have only positive (or negative) sensitivity [34]. The maximum sensitivity is concentrated close to the surface of the forearm, near the artery.

5. Transfer function

The electrical bioimpedance-based method for central aortic pressure waveform reconstruction allows long-term monitoring of the CAP and obtaining of haemodynamic parameters like the augmentation index (AI) [41] (see **Figure 15**).

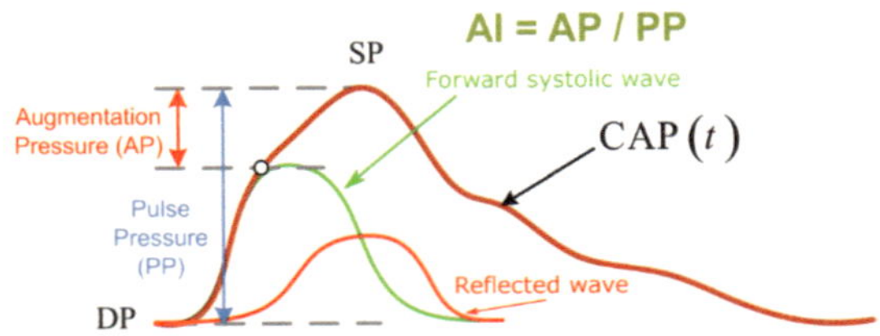

Figure 15.
Finding of augmentation index (AI) from the CAP waveform with diastolic and systolic pressures DP and SP accordingly, the CAP waveform is a sum of the forward (green line) and reflected (light red) pressure waves.

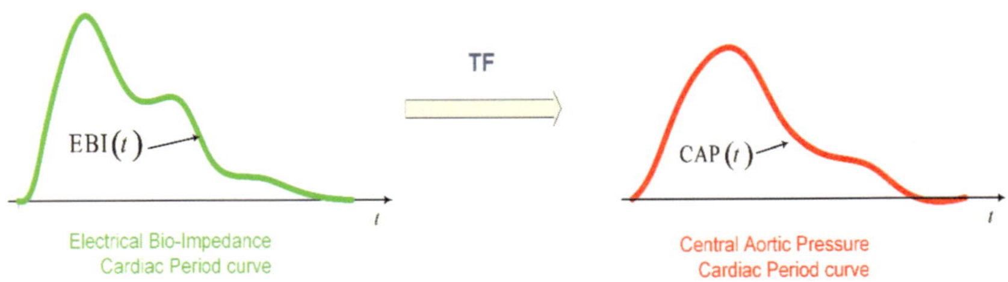

Figure 16.
Demonstration of the CAP cycle reconstruction with transfer function (TF) from the EBI cardiac cycle waveform.

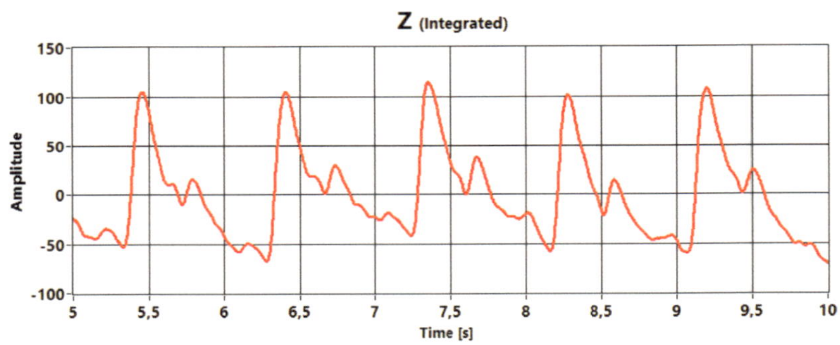

Figure 17.
*The impedance variation $\Delta Z(t)$, obtained by integration of $dZ(t)/dt$ wave (**Figure 7**) using trapezoidal method with dt = 0.0025 s and scaling the amplitude by factor of 20.*

To make this possible, the measured EBI waveform is transformed into the waveform of CAP, and for that, the transfer function (TF) approach is used (see **Figure 16** for illustration). Over 100 measurements of the EBI and invasive CAP waveforms were provided in East-Tallinn Central Hospital (Estonia) to collect data for this research work. In the beginning, the EBI measurements were carried out using a wireless multichannel impedance cardiograph CircMon BT101 [22] with additional channel for simultaneous acquiring of invasively measured CAP data using the PVB's XTRANS pressure sensor.

One possible algorithm for estimating a generic TF between the EBI and CAP waveforms (**Figures 17** and **18**) is a period-wise estimation of individual transfer functions for each patient and ensemble averaging to get a generic TF [42]. Another approach is to use adaptive algorithm to find a generic TF directly by matching all available patients' signals [43]. Both approaches give similar generic TF between the EBI and CAP of cardiac cycle waveforms. Despite the efforts made, the problems related to removing the artefacts remain. This causes corruption of the reconstructed CAP waveform due to the fact that all uncleaned artefacts are transformed into the reconstructed CAP. Regardless of that, the EBI signal-based noninvasive

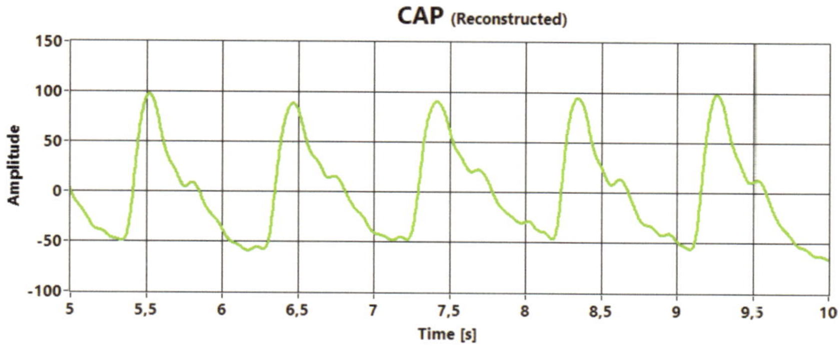

Figure 18.
The CAP waveform reconstructed digitally from the radial artery impedance waveform ΔZ(t) given in ***Figure 17*** *by applying the transfer function TF (**Figure 16**).*

estimation of the central aortic blood pressure waveform is still highly promising alternative to the applanation method.

6. Summary

Blood pressure variations inside the aorta during cardiac cycles (known also as central aortic blood pressure curve) is an important source for diagnosing patient's cardiovascular system. Classical approach, catheterization, is technically demanding and costly medical procedure. Therefore, different noninvasive methods have been studied and taken into use. The present chapter discusses the possibilities to bring in wearable techniques by sensing blood pressure variations with electrical bioimpedance changes on the radial artery. Two versions of wearable devices were designed and different electrode systems studied by simulations and experiments. Provisional human experiments were carried out at the hospital in limited extent but evidently with positive results. Further work will concentrate in developing the generalized transfer function algorithms and electrode system.

Conflict of interests

The authors declare no conflicts of interests.

Acknowledgements

The research was supported by Estonian ICT Center of Research Excellence EXCITE, Estonian Research Council (grant IUT1911), European project H2020 Flag-ERA JTC 2016 CONVERGENCE and Estonian IT Academy scholarships. The work has received funding from the European Union's Horizon 2020 research and innovation programme under grant agreement No 668995. This material reflects only the authors' view, and the EC Research Executive Agency is not responsible for any use that may be made of the information it contains.

Author details

Mart Min*, Hip Kõiv, Eiko Priidel, Ksenija Pesti and Paul Annus
Thomas Johann Seebeck Department of Electronics, Tallinn University of Technology, Tallinn, Estonia

*Address all correspondence to: mart.min@ttu.ee

References

[1] Chobanian A, Bakris G, Black H. The seventh report of the Joint National Committee on prevention, detection, evaluation, and treatment of high blood pressure. The JNC 7 report. ACC Current Journal Review. 2003;**12**(4):31-32

[2] McEniery CM, Cockcroft JR, Roman MJ, Franklin SS, Wilkinson IB. Central blood pressure: Current evidence and clinical importance. European Heart Journal. 2014;**35**(26):1719-1725

[3] Roman MJ, Devereux RB, Kizer JR, Lee ET, Galloway JM, Ali T, et al. Central pressure more strongly relates to vascular disease and outcome than does brachial pressure: The strong heart study. Hypertension. 2007;**50**(1):197-203

[4] Sharman J, Marwick T. Accuracy of blood pressure monitoring devices: a critical need for improvement that could resolve discrepancy in hypertension guidelines. Journal of Human Hypertension. 2018;**33**(2):89-93

[5] Picone DS, Schultz MG, Otahal P, Aakhus S, Al-Jumaily AM, Black JA, et al. Accuracy of cuff-measured blood pressure: Systematic reviews and meta-analyses. Journal of the American College of Cardiology. 2017;**70**(5):572-586

[6] Esper SA, Pinsky MR. Arterial waveform analysis. Best Practice & Research: Clinical Anaesthesiology. 2014;**28**(4):363-380

[7] Tavakol M, Ashraf S, Brener SJ. Risks and complications of coronary angiography: A comprehensive review. Global Journal of Health Science. 2012;**4**(1):65

[8] Karamanoglu M, O'rourke M, Avolio A, Kelly R. An analysis of the relationship between central aortic and peripheral upper limb pressure waves in man. European Heart Journal. 1993;**14**(2):160-167

[9] Avolio AP, Butlin M, Walsh A. Arterial blood pressure measurement and pulse wave analysis—Their role in enhancing cardiovascular assessment. Physiological Measurement. 2009;**31**(1):R1

[10] Cloud GC, Rajkumar C, Kooner J, Cooke J, Bulpitt CJ. Estimation of central aortic pressure by SphygmoCor requires intra-arterial peripheral pressures. Clinical Science. 2003;**105**(2):219-225

[11] Butlin M, Qasem A. Large artery stiffness assessment using SphygmoCor technology. Pulse. 2016;**4**(4):180-192

[12] Miyashita H. Clinical assessment of central blood pressure. Current Hypertension Reviews. 2012;**8**(2):80-90

[13] Chung E, Chen G, Alexander B, Cannesson M. Non-invasive continuous blood pressure monitoring: A review of current applications. Frontiers in Medicine. 2013;**7**(1):91-101

[14] Omboni S, Posokhov IN, Kotovskaya YV, Protogerou AD, Blacher J. Twenty-four-hour ambulatory pulse wave analysis in hypertension management: Current evidence and perspectives. Current Hypertension Reports. 2016;**18**(10):72

[15] Harju J, Vehkaoja A, Kumpulainen P, Campadello S, Lindroos V, Yli-Hankala A, et al. Comparison of non-invasive blood pressure monitoring using modified arterial applanation tonometry with intra-arterial measurement. Journal of Clinical Monitoring and Computing. 2018;**32**(1):13-22

[16] Herscovici H, Roller DH. Noninvasive determination of central blood pressure by impedance plethysmography. IEEE Transactions on Biomedical Engineering. 1986;**6**:617-625

[17] Hatschek RA. Blood pressure measuring device and method. Google Patents; 1994

[18] Matsumura N, Sawanoi Y, Iwahori T. Pulse wave measurement electrode unit and pulse wave measurement device. Google Patents; 2010

[19] Solà J, Adler A, Santos A, Tusman G, Sipmann FS, Bohm SH. Non-invasive monitoring of central blood pressure by electrical impedance tomography: First experimental evidence. Medical and Biological Engineering and Computing. 2011;**49**(4):409

[20] He J, Wang M, Li X, Li G, Lin L. Pulse wave detection method based on the bio-impedance of the wrist. Review of Scientific Instruments. 2016;**87**(5):055001

[21] Cohn GA, Kusche R. Bioimpedance based pulse waveform sensing. Google Patents; 2018

[22] Annus P, Lamp J, Min M, Paavle T. Design of a bioimpedance measurement system using direct carrier compensation. In: Proceedings of the 2005 European Conference on Circuit Theory and Design. 2005. pp. 3-23

[23] Priidel E, Annus P, Metshein M, Land R, Märtens O, Min M. Lock-in integration for detection of tiny bioimpedance variations. In: 2018 16th Biennial Baltic Electronics Conference (BEC). 2018. pp. 1-4

[24] Krivoshei A, Lamp J, Min M, Uuetoa T, Uuetoa H, Annus P. Non-invasive method for the aortic blood pressure waveform estimation using the measured radial EBI. Journal of Physics: Conference Series. 2013;**434**:012048

[25] Grimnes S, Martinsen O. Bioelectricity and bioimpedance basics. London: Academic; 2008

[26] Tallgren P, Vanhatalo S, Kaila K, Voipio J. Evaluation of commercially available electrodes and gels for recording of slow EEG potentials. Clinical Neurophysiology. 2005;**116**(4):799-806

[27] Lu F, Wang C, Zhao R, Du L, Fang Z, Guo X, et al. Review of stratum corneum impedance measurement in non-invasive penetration application. Biosensors. 2018;**8**(2):31

[28] Metshein M. Wearable solutions for monitoring cardiorespiratory activity [PhD thesis]. Tallinn University of Technology; 2018

[29] Jung H-C, Moon J-H, Baek D-H, Lee J-H, Choi Y-Y, Hong J-S, et al. CNT/PDMS composite flexible dry electrodesfor long-term ECG monitoring. IEEE Transactions on Biomedical Engineering. 2012;**59**(5):1472-1479

[30] Kõiv H, Pesti K, Min M, Land R. Investigation of Cost-Effective Carbon Nanofiber/Carbon Fiber and Silicone Polymer Composite Material for Wearable Bioimpedance Device. 2019. Forthcoming

[31] Metshein M, Kõiv H, Annus P, Min M. Electrode optimization for bio-impedance based central aortic blood pressure estimation. In: World Congress on Medical Physics and Biomedical Engineering. 2018. pp. 497-501

[32] Lazovic B, Mazic S, Zikich D, Zikic D. The mathematical model of the radial artery blood pressure waveform through monitoring of the age-related changes. Wave Motion. 2015;**56**:14-21

[33] Kauppinen P, Hyttinen J, Malmivuo J. Sensitivity distribution visualizations of impedance tomography measurement strategies. International Journal of Bioelectromagnetism. 2006;**8**(1):1-9

[34] Pesti K, Kõiv H, Min M. Simulation of the Sensitivity Distribution of Four-Electrode Impedance Sensing on Radial Artery. 2019. Forthcoming

[35] Canali C, Mazzoni C, Larsen LB, Heiskanen A, Martinsen OG, Wolff A, et al. An impedance method for spatial sensing of 3D cell constructs-towards applications in tissue engineering. Analyst. 2015;**140**(17):6079-6088

[36] Geselowitz DB. An application of electrocardiographic lead theory to impedance plethysmography. IEEE Transactions on Biomedical Engineering. 1971;**1**:38-41

[37] Brown BH, Seagar AD. The Sheffield data collection system. Clinical Physics and Physiological Measurement. 1987;**8**(4A):91

[38] Hua P. Effect of the measurement method on noise handling and image quality of EIT imaging. In: Proc Annu Int Conf Engng Med and Biol Soc. 1987. pp. 1429-1430

[39] Holder DS. Electrical impedance tomography. Bristol: IoP Publishing; 2005

[40] Islam N, Rabbani KS, Wilson A. The sensitivity of focused electrical impedance measurements. Physiological Measurement. 2010;**31**(8):S97

[41] Krivoshei A, Min M, Lamp J, Annus P. Robust algorithm for the augmentation index estimation of the CAP using low order derivatives. In: 2014 IEEE International Symposium on Medical Measurements and Applications (MeMeA). 2014. pp. 1-4

[42] Krivoshei A, Min M, Uuetoa H, Lamp J, Annus P. Electrical bio-impedance based non-invasive method for the central aortic blood pressure waveform estimation. In: 2014 14th Biennial Baltic Electronic Conference (BEC). 2014. pp. 181-184

[43] Min M, Annus P, Kõiv H, Krivoshei A, Uuetoa T, Lamp J. Bioimpedance sensing-a viable alternative for tonometry in non-invasive assessment of central aortic pressure. In: 2017 IEEE International Symposium on Medical Measurements and Applications (MeMeA). 2017. pp. 373-378

Chapter 3

Wearable Skin-Worn Enzyme-Based Electrochemical Devices: Biosensing, Energy Harvesting, and Self-Powered Sensing

Itthipon Jeerapan

Abstract

Integrating enzymes with wearable electrochemical systems delivers extraordinary functional devices, including biosensors and biofuel cells (BFCs). Strategies employing enzyme-based bioelectronics represent a unique foundation of wearables because of specific enzyme recognition and catalytic activities. Therefore, such electrochemical biodevices on various platforms, e.g., tattoos, textiles, and wearable accessories, are interesting. However, these devices need effective power sources, requiring combining effective energy sources, such as BFCs, onto compact and conformal platforms. Advantageously, bioenergy-harvesting BFCs can also act as self-powered sensors, simplifying wearable systems. Challenges pertaining to energy requirements and the integration of biocatalysts with electrodes should be considered. In this chapter, we detail updated advancement in skin-worn devices, including biosensors, BFCs, and self-powered sensors, along with engineering designs and on-skin iontophoretic strategies to extract biofluids. Crucial parameters including mechanical/material aspects (e.g., stretchability), electrochemistry, enzyme-related views (e.g., electron shuttles, immobilization, and behaviors), and oxygen dependency will be discussed, along with outlooks. Understanding such challenges and opportunities is important to revolutionize wearable devices for diverse applications.

Keywords: wearable technology, electrochemical devices, enzyme-based bioelectronics, biosensors, biofuel cells, self-powered biosensors, sweat, iontophoresis, personalized healthcare

1. Introduction

Since 1962 when the first Clark's biosensor was introduced [1], enzymatic electrochemical devices have attracted increasing attention, recently being regarded as a powerful tool for the development of emerging wearable bioelectronics [2]. Integrating enzymes with electrochemical transduction units is one of the most popular and well-built bioelectronic systems due to outstanding selectivity and natural behaviors of enzymes [2–4]. Employing enzymes, as a catalytic system, in order to substitute nonselective metal catalysts, is interesting. Because of inherent behaviors of enzymes, enzyme-based bioelectronics offers favorable operations under mild physiological conditions of pH and temperature, unlike nonenzymatic approaches [5, 6]. In addition, enzymes will usually catalyze only one particular reaction. Therefore, such enzyme specificity enables bioelectronics to operate selectively even in complex solutions, including biofluids. Recently, there is an increasing interest in transforming traditional enzymatic bioelectronics into modern wearable platforms. Wearable enzyme electronics expands appealing spectra of a variety of applicable fields, ranging from personalized healthcare, fitness, to the environment. These applications comprise of noninvasive diagnosis of biomarkers in biofluids, such as sweat, and the monitoring of the surrounding of the wearer. Besides, electron collectors can be functionalized with enzymes to develop BFCs for energy and self-powered applications. These biodevices employ enzymes to obtain electrocatalytic oxidations of biofuels, such as glucose and lactate. This aims to achieve next-generation energy autonomy for the whole wearable system. In addition to energy-harvesting purposes, BFCs can also act as self-powered electrochemical sensors. Three main applications of enzyme-based

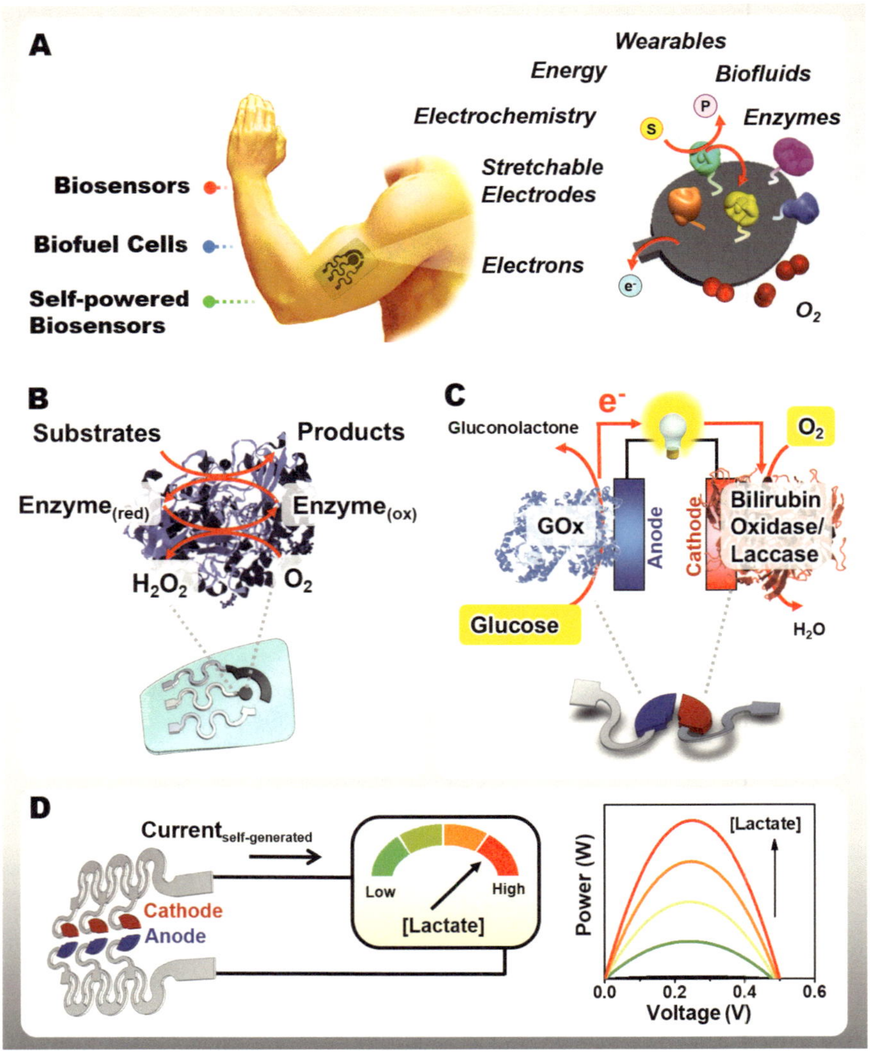

Figure 1.
(A) Skin-worn enzyme-based electrochemical devices. The soft electrode platform is functionalized with enzymes, allowing various applications, including (B) biosensors, (C) energy-harvesting biofuel cells, and (D) self-powered biosensors.

electrodes, including biosensors, biofuel cells (BFCs), and self-powered sensors, along with their relevant aspects, will be discussed (**Figure 1**). An enzymatic biosensor employs an enzyme, immobilized on an electrochemical transducer, to recognize and react with the target, generating a readable electrical signal (**Figure 1B**). A BFC energy harvester can convert chemical energy into electricity and power wearable devices (**Figure 1C**) [7]. A BFC can also be designed to act as a self-powered sensor by displaying power signals proportional to the target concentration (**Figure 1D**) [8, 9].

Skin-worn enzyme-based electrochemical devices are among the most significant wearables because the skin offers the largest organ interface and unique opportunities to be accessed noninvasively [10–13]. The large epidermal area also provides sweat, which contains a variety of biomarker-rich information, such as levels of glucose, lactate, hormone, urea, pH, and electrolytes. Advantageously, skin-worn electrochemical devices can be attached directly close to the location of sweat generation, enabling the fast access for monitoring or energy harvesting before the unwanted biodegradation. In addition to physical parameters obtained from existing skin-worn biodevices (such as temperature and heartbeat), chemical data is also crucial to step further to understand comprehensive insights of individual [14]. The history of sweat content analysis began many decades ago with the development of cystic fibrosis diagnosis [15]. Establishing new "lab-on-skin" electrochemical devices enables noninvasive detection of such biometrics, essential for health monitoring and early disease diagnosis. In addition, such wearable electrochemical tools are also helpful for drug testing and chemical threat screening, such as in sports [12] and in the surrounding environment [16]. Importantly, for emerging energy technologies, sweat also contains relevant biofuels, such as glucose and lactate; this is useful to BFCs as energy-harvesting and self-powered devices, which exemplify new exciting wearable autonomous bioelectronic systems.

Although researchers are battling to create new enzymatic bioelectronics, there is a continuing need for further development. Revolutionizing traditional electrodes toward wearable bioelectronics needs careful engineering to address several key challenges associated with electrochemistry, the integration of biocatalysts, mechanical stability, environment effects (e.g., O_2 fluctuations), and sweat extraction. Therefore, the bulk of this chapter will focus on examples of progress in skin-worn enzymatic electrochemical devices. Key working principles and opportunities of biosensors and BFCs will be described. In addition, perspectives emphasizing on main challenges will be discussed. The outlooks of emerging wearable electrochemical technologies will also be concluded.

2. Skin-worn enzyme-based electrochemical devices

2.1 Enzyme-based biosensors

Wearable enzymatic electrochemical biosensors utilize enzymes, which are functionalized in spatial contact with electrochemical transduction units. In principle, biosensors consist of electrodes and enzyme receptors, allowing the specific binding capabilities and catalytic activity to target analytes. Interfacing enzymes with electrodes will be discussed further in Section 3.3. It should be remarked that the key consideration to fabricate a successful biosensor for nonspecialist wearers is choosing highly specific biocatalysts. Enzymatic biosensors can also function continuously because enzymes are not consumed in reactions, offering an advantage for wearable sensors.

Enzymatic biosensors are based on numerous mechanisms. The popular mechanism relies on the conversion of the analyte as an enzymatic substrate into a product, enabling the detection by using electrochemical transducer. Another way is to monitor the analyte (e.g., a toxic compound) that acts as an enzyme inhibitor. In addition, the enzyme can be used as a labeling transducer for bioaffinity recognition. Besides, a reverse approach can be designed to detect the enzyme level. In this case, the enzyme acts as an analyte, while the substrate is immobilized on the electrode surface. When the enzyme reaches the electrode sensor, it will generate the signal, corresponding to the concentration level of the enzyme target.

In recent decades, enzymatic biosensors have been proven to be modern wearables to monitor numerous analytes, such as glucose, lactate, alcohol, and organophosphate nerve agents. Among several enzymes, oxidoreductase and hydrolase, such as glucose oxidase (GOx), lactate oxidase (LOx), alcohol oxidase (AOx), and organophosphorus hydrolase, are predominant for wearable biosensing applications. A temporary tattoo with the integration of transdermal enzymatic glucose biosensor has been introduced since glucose is a key biomarker for diabetes mellitus, which still affects hundreds of millions of patients globally (**Figure 2A**) [17]. The iontophoretic ISF extraction system was coupled with the amperometric

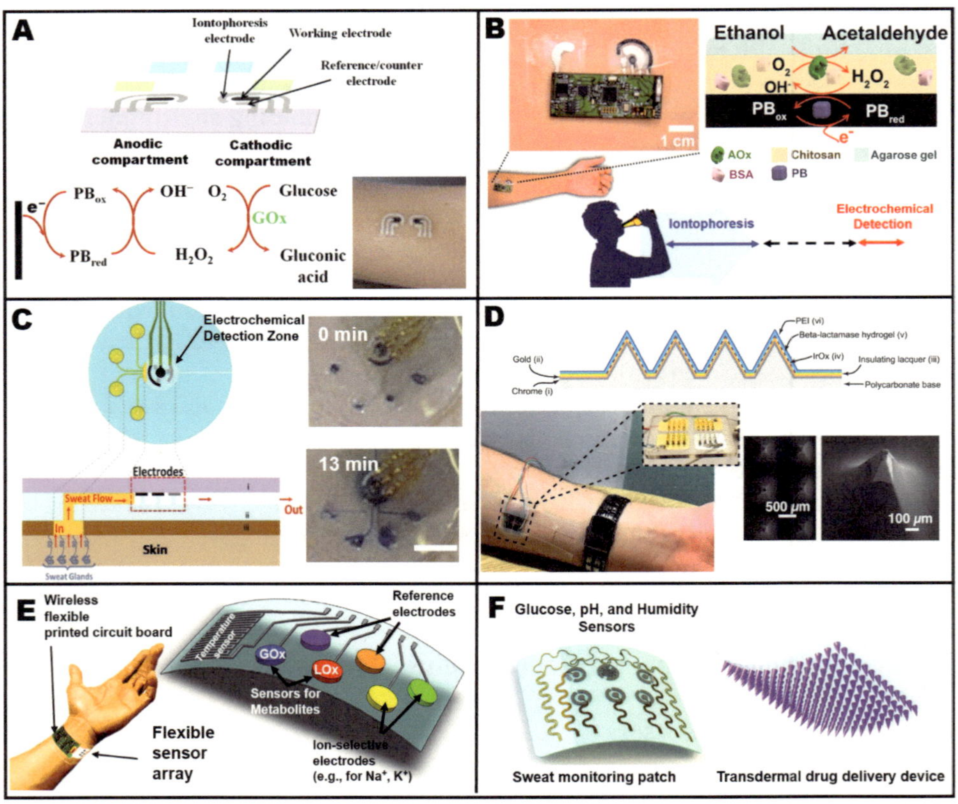

Figure 2.
Skin-worn enzyme-based electrochemical biosensors. (A) Transdermal tattoo-based glucose sensors, coupled with reversed iontophoresis [17]. Adapted with permission from ref [17]. Copyright 2015 American Chemical Society. (B) Tattoo-based alcohol biosensors, coupled with pilocarpine iontophoresis and wireless electronics [18]. Adapted with permission from ref [18]. Copyright 2016 American Chemical Society. (C) Biosensors integrated with a microfluidic patch for sweat collection and analysis [20]. Adapted with permission from ref [20]. Copyright 2017 American Chemical Society. (D) Microneedle-based β-lactam sensors [22]. Adapted with permission from ref [22]. Copyright 2019 American Chemical Society. (E) Integrated glucose/lactate enzymatic biosensors with electrolyte and temperature sensors. (F) Integrated sweat monitoring biosensing and transdermal drug delivery system.

detection to extract the sample containing glucose. The glucose biosensor, located near the negative iontophoretic electrode, relied on GOx immobilization on the Prussian blue (PB)-carbon electrode; this PB facilitates the electroreduction of H_2O_2 product, generated by the GOx reaction. The amperometric reduction of H_2O_2 could be detected at a potential of −0.1 V versus Ag/AgCl. The iontophoresis strategy will be discussed in Section 3.5. Additionally, the tattoo-based alcohol sensor was also invented (**Figure 2B**). The AOx-/PB-based sensor was designed to be close to the positive iontophoretic electrode to determine ethanol in sweat induced by transdermal delivery of the pilocarpine drug [18]. Moreover, recent efforts have been made to combine these two concepts, including glucose and alcohol sensors, on a single tattoo [19]. This holds a possibility for multianalyte sweat analysis.

Skin-worn microfluidic devices can enable the continuous flow of renewed sweat over operational periods. This addresses the challenge of mixing and carry-over between new and old sweat. **Figure 2C** shows an example of sweat collection microfluidic devices, coupled with glucose and lactate biosensors [20]. This offers wearable effective continuous sweat sampling and flow electroanalysis.

Furthermore, minimally invasive microneedles for continuous glucose monitoring have been demonstrated. For example, a GOx/tetrathiafulvalene microneedlebased amperometric sensor (~1.2 mm needle height) could be used for in vivo studies [21]. The data were also validated with the finger-prick technique, indicating a promising alternative for on-skin analysis. In addition, a minimally-invasive microneedle-based potentiometric sensor for tracking β-lactam antibiotic concentrations in vivo and real time was demonstrated **Figure 2C** [22]. This example represents a possibility to tailor individual therapy with the optimal efficacy.

Moreover, reading several parameters can complete a clear picture of individual health. A fully integrated sensor array for sweat analysis was demonstrated (**Figure 2E**) [23]. These integrated sensors can monitor information of glucose, lactate, electrolytes (e.g., sodium and potassium ions), and temperature. The temperature sensor is also helpful to standardize the biosensing amperometric response. Furthermore, in order to apply the biosensor glucose device for health management, a transdermal closed-loop drug delivery integrated with a sweat-based glucose electrochemical sensor was demonstrated (**Figure 2F**) [24]. The sense-treat concept aimed to give feedback of transdermal administration of type 2 diabetes drugs in response to the glucose level. This idea represents a possible opportunity to overcome insulin overtreatment, helping patients to maintain their homeostasis.

2.2 Enzyme-based electrochemical power sources

BFCs are energy-conversion devices that utilize biocatalysts to convert chemical energy into electricity. For wearable electronics, the need to anatomically power sources has attracted many research groups to develop a BFC, as a "green" energy-harvesting alternative, in order to extract energy from metabolites present in biofluids, such as perspiration. Since glucose, lactate, and oxygen are present in physiological fluids, in general, a majority of wearable enzymatic BFCs rely on (1) the generation of electrons from glucose or lactate biofuels and (2) the electron reduction by oxidants (such as oxygen). **Figure 1C** shows a typical example of a glucose/O_2 BFC. In principle, a glucose BFC uses GOx, functionalized on the bioanode, to catalyze the glucose oxidation reaction to generate electrons. After this oxidation process, these harvested electrons are driven through an external circuit to the biocathode compartment where such electrons are accepted by oxidant molecule (commonly O_2) and, eventually, generate complete electrical work. In addition to

Pt-based catalysts, multicopper oxidases such as laccase, bilirubin oxidase, and polyphenol oxidase are commonly used for electrocatalyzing oxygen-reduction reaction (ORR) in the BFC cathode [25].

Enzymatic BFCs represent an interesting alternative due to their unique advantages, such as outstanding selectivity and behaviors of enzymes. Unlike most traditional inorganic catalyst-based fuel cells, which require harsh conditions (such as acidic conditions or high temperatures ranging from 45°C to more than 100°C), the enzyme-based BFC can operate under mild conditions (20–40°C at neutral pH). Moreover, non-specific catalyst-based fuel cells require to separate anode and cathode chambers by a thin membrane. Unfortunately, this common use of separation membrane between the anode and the cathode compartments will be unsatisfactory for skin-worn miniaturized devices. Thanks to the nature of enzymes, utilizing high specificity of enzymatic catalysis can obviate this membrane requirement, facilitating the fabrication and applications [26]. In addition, enzyme-based BFCs can operate selectively in complex biofluids.

Interestingly, BFCs also offer opportunities to design self-powered biosensors (**Figure 1D**). For example, the power is proportional to the concentration of the fuel (also acting as analyte); self-powered output itself can determine the level of the target. This offers opportunities to eliminate external energy sources for powering potentiostat and signaling systems [9].

An initial concept integrating enzymatic BFCs with skin-worn technologies represented an exciting way to scavenge bioenergy available in human perspiration (**Figure 3A**). This demonstrated the first epidermal tattoo-based BFC that converted sweat lactate biofuel and oxygen into electricity [27]. The lactate oxidation by LOx electrocatalyzation was mediated by tetrathiafulvalene on the carbon nanotube (CNT)-based anode, while electroreduction on the oxygen-reduction cathode relies on Pt black catalyst. This system facilitates mediated oxidation of lactate at −0.1 V with a peak potential of 0.14 V (versus Ag/AgCl). This low anodic onset potential indicates the efficient electron-donor-acceptor TTF/CNT. The successful on-body test displayed a power up to 70 μW cm^{-2}. This idea was also established on fabrics and could power a light-emitting diode with an integrated DC-DC converter [28].

Mechanical stability has been the focus in the development of the next-generation of skin-worn BFCs due to the multiplex mechanical movements experienced *in vivo*. In order to minimize cracking of the device and maintain good electrochemical performance, screen-printable stretchable inks and judicious stretchable design have been engineered (**Figure 3B**) [29]. Combining additional degrees of stretchability with intrinsic mechanical resiliency of soft CNT/polyurethane (PU) composite offers the desirable stretchable platform. The BFC was then functionalized on the soft electrodes, allowing good mechanical stability. This holds promise applications for on-body bioenergy fields wherein resilience toward mechanical distortions is compulsory.

In addition to energy-conversion applications, BFCs can be applied further as another significant tool for wearable bioelectronics. Enzymatic BFC can serve as self-sustainable biosensors (without an extra powering device). In order to expand the spectrum of BFC applications for on-skin electroanalytical chemistry, the pioneering stretchable textile-based BFCs that can act as self-powered was demonstrated (**Figure 3C**) [30]. These biodevices can deliver two key functions: (1) harvesting electrical power from sweat glucose and lactate and (2) displaying signals of such metabolites. Extracted bioenergy from the wearer's sweat can directly indicate the metabolite levels. Sock-based biodevices were successfully demonstrated on human subjects, representing a promising concept for modern wearable self-powered biosensors.

Maximizing the loading amount of active enzyme, mediator, and conductive materials can improve the power performance of BFCs. The high amount of such active materials can be packed by a compress. However, this strategy will affect mechanical softness. Therefore, further engineering was to fabricate island-bridge assemblies merging the high enzyme loading packed islands with stretchable serpentine bridges [34]. This combination offered a soft bioelectronic skin for harvesting a good power density of 1.2 mW cm^{-2}. This energy was sufficient to power a Bluetooth Low Energy (BLE) radio integrated with a DC-DC converter.

Recently, additional efforts have been made to scavenge, improve, and store energy by hybridizing textile-based energy conversion with energy storage devices (BFCs and supercapacitors, respectively) (**Figure 3E**) [31]. The on-body demonstration showed that after perspiring, the supercapacitor could be charged by the BFC energy and reach a stable 0.4 V output.

Furthermore, a photoelectric BFC was developed to convert external light and-chemical energy from wearer's perspiration into electrical energy (**Figure 3D**) [32]. The anode relied on a LOx/Meldola's blue/buckypaper electrode, while the photocathode

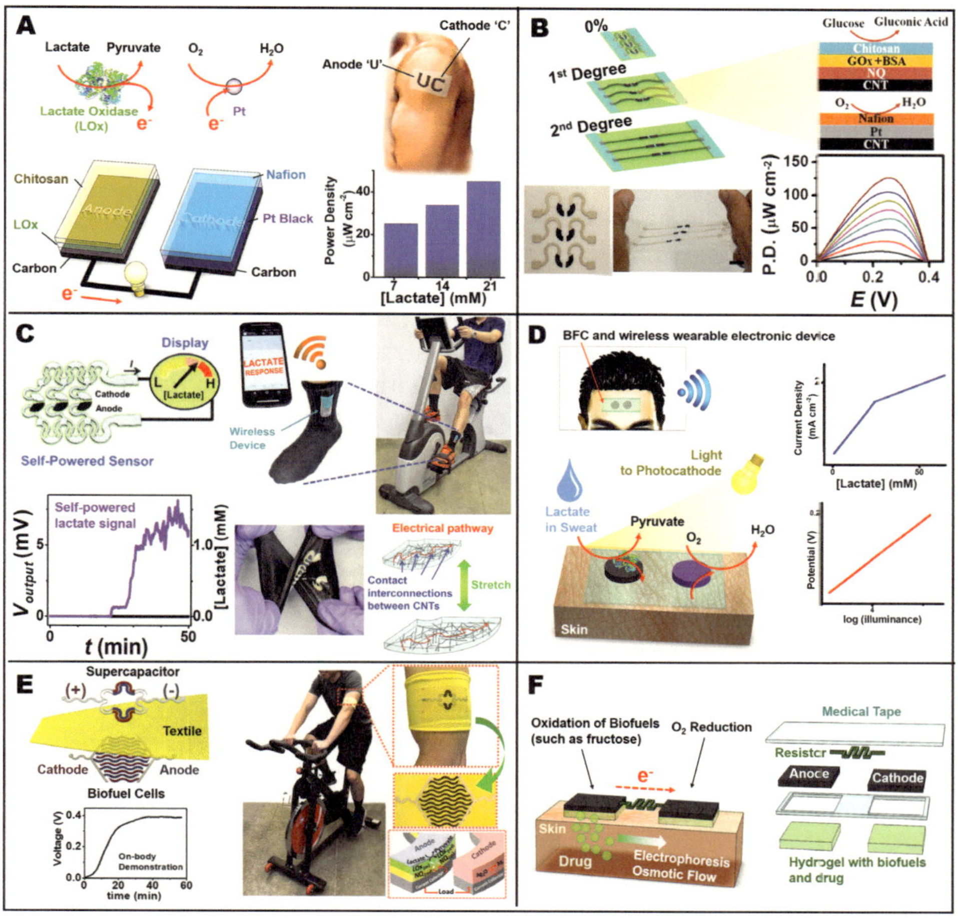

Figure 3.
Skin-worn BFCs and self-powered sensors. (A) Epidermal tattoo-based lactate BFCs. (B) Stretchable glucose BFCs [29]. Adapted with permission from ref [29]. Copyright 2016 American Chemical Society. (C) Stretchable textile-based BFCs acting as self-powered biosensors [30]. Adapted with permission from ref [30]. Copyright 2016 The Royal Society of Chemistry. (D) Photoelectric BFCs. (E) Textile-based BFC-supercapacitor hybrid devices [31]. Adapted with permission from ref [31]. Copyright 2018 The Royal Society of Chemistry. (F) Built-in BFCs with transdermal iontophoresis patches.

relied on an organic polyterthiophene semiconductor, which drove a reduction reaction under illumination (wavelengths of 350 nm to over 600 nm). This system presented an attractive example of on-skin autonomous power sources and sensors.

Additional efforts have been made to explore new biomedical applications of BFCs. **Figure 3F** shows an integrated fructose/O_2 BFC patch that was conjugated with transdermal iontophoresis [33]. The current generated by the BFC was used to drive an osmotic flow from the anode to the cathode, resulting in the net ionic movement of small-molecule drug into the skin. The level of transdermal current to control the drug administration could be adjusted by connecting a thin poly(3,4-ethylenedioxythiophene)/PU resistor of a programmable resistance value.

3. Challenges and possible solutions

3.1 Mechanical properties

Young's modulus of the human skin is in a range of 10–500 kPa [35, 36], while the moduli of common electronic materials, such as silicon and gold, are much higher (high GPa), indicating significant mechanical mismatch when integrating with the skin. Therefore, functionalities of non-stretchable electrodes will deteriorate after multiplex deformations commonly experienced by daily life activities. Furthermore, such rigidity and bulkiness of traditional devices also restrict the wearability and comfortability [14]. Non-compliant electrochemical devices will limit continuous long-term functions due to cracking and increasing of material resistance. This increasing of resistivity, which opposes the current flow in bioelectronics, causes poor electron communication at the enzyme-electrode interface.

This major challenge of skin-integrated electronics can be addressed by exploring stretchable materials which display mechanical properties in a similar range of skin's modulus. One approach is using polymers due to their low mechanical toughness. For example, conducting materials with high moduli can be blended with soft polydimethylsiloxane or Ecoflex materials (Young's moduli of 0.4–3.5 MPa and 125 kPa, respectively) in order to tune the mechanical properties while keeping good electrochemical functions [37]. CNT-based materials, which are powerful for electrochemical devices [38], are used to combine with soft elastomers, such as PU and styrene-butadiene-styrene (SBS) [29, 39]. PU and SBS composites have moduli of ~700–800 kPa. As shown in **Figure 3C**, CNT filler (with the high-aspect ratio ~1300) was combined with PU [30], achieving stretchable conductive electrode materials. The percolation of dispersed CNTs can facilitate the electric flow in stretchable bioelectronics. Combining the intrinsic stretchability of this engineered inks with the structural stretchability of the serpentine design allows the device to tolerate strains as high as 500% with a small effect on its electrochemical performance [29]. This concept can be expanded by adding new functionalities into electrodes. For example, platinum-decorated graphite was mixed with PU to obtain stretchable electrocatalytic materials, allowing the fabrication of stretchable electrodes for glucose biosensors [40].

3.2 Powering wearable devices

Growing demand of wearable technologies has stimulated the need of the development of viable energy sources. The lack of anatomically power sources becomes a key bottleneck for the progress in wearable bioelectronics. Skin-worn bioelectronics mandates the compliant and efficient energy sources to supply multitasks, including sensing and data communication. In addition to developing low-power-consuming electronic microelectronics [9, 41], there is an increasing interest in advancing bioenergy-harvesting devices. Enzymatic BFCs are attractive self-sustainable energy devices to meet this growing energy demand. For example, 0.3-V complementary metal-oxide-semiconductor (CMOS) wireless glucose or lactate biosensing systems, which consumed power of ~1.2 μW, could be powered by BFCs [9]. Nevertheless, several applications of enzymatic BFCs still have some challenges, such as low-power output. The major challenge in enzymatic BFC is faced by the electrical "wiring" of enzymes with electrodes. The difficulty of electrical wiring, referring to electron transfer, and their possible solutions will be detailed in Section 3.3.

Compared with traditional fuel cells, enzymatic BFCs are challenging due to their multicomponent including redox potentials of enzyme, cofactor, and mediator. This results in the typical unwanted deviation of open-circuit voltages (OCV) from their theoretical maximum values, referring to "cell voltage losses." The redox potential for electrocatalytic oxidation at the bioanode required to be higher than that of the biocathode for reduction reaction in order to deliver a sufficient electromotive force for electron transfer between enzyme active site and mediator. The voltage difference between the formal redox potentials ($E^{\circ\prime}$) of redox enzyme cofactors in the active sites, in the anode and cathode, will govern the maximum cell voltage. Parameters, including redox potential of mediator and cofactor redox potential in the enzyme, can influence the resulting potential output of BFCs. Therefore, the mediator should be carefully chosen. For example, ferrocene derivatives coimmobilized with GOx at a graphite electrode can be used for glucose sensors [42]. Nevertheless, ferrocene derivatives display high redox potentials (0.1–0.4 V versus SCE); these will cause cell voltage losses in the GOx-based BFC if they are used as anode mediators. It should be noted that the difference between the redox potentials of the enzymes wired at the anode and the cathode determines the cell voltage. An example of a successful anode mediator used in skin-worn BFCs is 1,4-naphthoquinone [30]. This quinone compound is also almost insoluble in cold water, preventing leaching during on-body operations. One challenge of using GOx on the anode is the O_2 competition with a mediator, decreasing the oxidation current on the bioanode. Moreover, O_2 competitive reaction on the anode can produce H_2O_2. This by-product can inhibit GOx activity and decrease the overall BFC performance. Therefore, catalase should be cofunctionalized to the bioanode to diminish the undesirable H_2O_2 [43].

A single-enzyme BFC can usually convert only a partial portion of biochemical energy, resulting in low current output. For instance, wearable BFCs, such as for harvesting energy from lactate sweat, commonly employ a single enzyme-based bioanode, catalyzing the oxidation of lactate to pyruvate, which only harvests two electrons. In other words, they utilize only a portion of the biofuel energy and leave most of the energy in the oxidized product. Therefore, it is interesting to harvest the total of 12 electrons in order to maximize the energy-conversion efficiency. A potential solution is to design an enzyme cascade system for complete oxidation of lactate fuel. For example, the bioinspired multienzyme catalytic cascade could complete the metabolic cycle, successfully enhancing net BFC power [44].

Furthermore, in order to optimize the current output, diffusion and enzyme loading should be enhanced. The engineering of specific enzyme activity and three-dimensional structure of enzymatic electrodes should be explored.

3.3 Enzyme-related aspects

The selection of enzymes is a primary subject which should be discussed. Enzymes must be selected by considering their particular reactions to target analytes or biofuels for electroanalytical monitoring and energy harvesting, respectively. One of the most predominant enzymes used to develop wearable bioelectronics is GOx from *Aspergillus niger*. It represents an example of commercially available biocatalyst that has good stability, substrate specificity, and electron turnover rate [3, 4, 45]. It is a powerful biorecognition element for glucose biosensors, the most widely interesting devices for diabetes health management. As shown in **Figure 4** (A–C), the enzyme is immobilized on the electrode, establishing a biosensor. GOx contains two 80 kDa subunits. Each holds a tightly bound flavin adenine dinucleotide (FAD) cofactor, the important redox center which has a redox potential −0.32 V (vs Ag/AgCl) at pH 7. This redox center is crucial to transfer electrons and specifically oxidize β-D-glucose to gluconolactone. However, this FAD is shielded by the protein and a glycan structure, hindering electron exchange at the enzyme-electrode interface. Inevitably, this requires research efforts to address this roadblock [46, 47]. FAD plays an important role as a common cofactor for glucose oxidation biocatalysis. The redox process for FAD/$FADH_2$, involving two electrons, is shown in **Figure 4D**, where the R group represents adenosine diphosphate and ribitol connected with the flavin. However, it is O_2-dependent; accordingly, O_2 fluctuations can vary the performance of this type of oxidase-based bioelectronics. Although alternative O_2-independent electrodes utilized NAD-dependent electrodes can be used, they require a diffusional cofactor, not simple for wearable applications. Hence, FAD-dependent dehydrogenases are becoming interesting choices since they are O_2-independent and do not depend on diffusional mediators [48, 49].

The first generation of biosensors relies on quantifying O_2 generation or H_2O_2 depletion (**Figure 4A**). This leads to key drawbacks, such as low dynamic range, dependency to oxygen fluctuations, and interfering effects. For instance, for glucose amperometric sensors, the detection of H_2O_2 at common first-generation electrodes needs the high applied detection potential where interfering compounds existing in sweat, e.g., ascorbic acid, uric acid, and some drugs, are also electroactive. Lowering the applied potential for the detection is a strategy to minimize

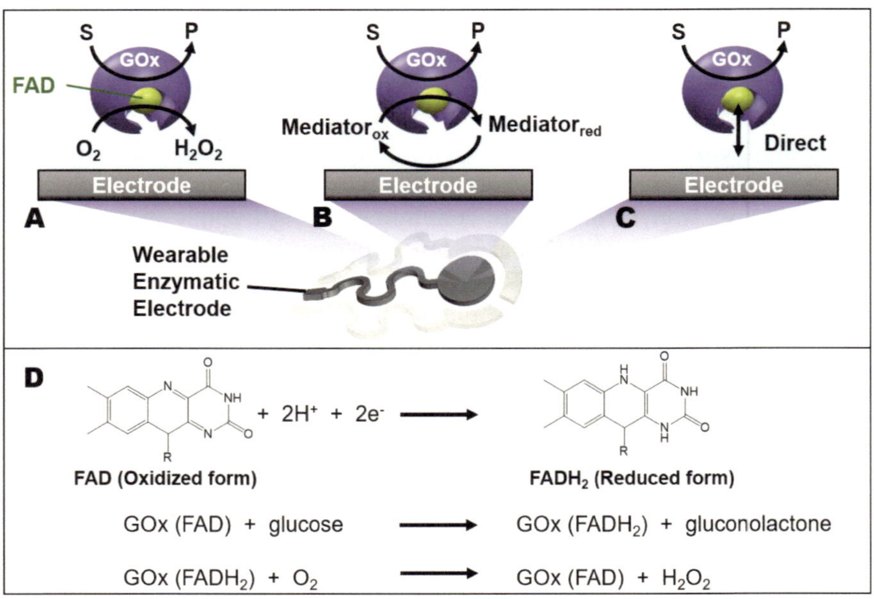

Figure 4.
Principles of interfacing the enzyme, such as glucose oxidase (GOx), with the electrode. Different generations of strategies (A–C: first, second, and third generations) are illustrated. (D) Reactions involving the glucose oxidase biocatalyst.

such electroactive interferences. One approach is to incorporate electrocatalysts in wearable electrodes, such as PB or Pt [17, 40]. This offers low-potential detection of H_2O_2 to mitigate interference effects.

Furthermore, researchers have developed two strategies to wire enzymes to the electrode interface (**Figure 4B** and **C**). These include (1) mediated electron transfer (MET) and (2) direct electron transfer (this may refer to mediatorless electron transfer between the enzyme and the electrode). Such new tactics are not only useful for enzymatic biosensors but also for enzymatic BFCs which also involve bioelectrocatalysis.

First, the MET strategy utilizes a redox mediator, acting as an electron-shuttle assistant between the enzymatic active center and the electrode. The substrate level, such as glucose, can then be monitored by the redox process of the mediator. This results in the independence of oxygen and mitigating the interfering signals due to the operation at low potentials. The first consideration in electrically wiring the enzyme with the electrode is the choice of the mediator that should be close to the redox potential of the active center of the enzyme to facilitate efficient electron communication between the enzyme and the conductive electrode surface. In particular, for enzymatic BFCs, the selection of mediators is crucial to positively control the cell voltage and enhance heterogeneous electron transfer to the order of a homogeneous transfer [50]. However, challenges of using mediators, particularly for BFCs, are their stability and deviated cell voltage. In addition, biocompatibility is highly vital for skin-worn applications. In spite of the assistance of electron shuttle by redox mediators, major concerns are their biocompatibility. One possible solution is employing nanomaterials or highly biocompatible catalysts. For example, mushroom/plant extracts could be used to obtain efficient "green" bioelectrocatalytic reactions for ethanol BFCs [51].

Second, direct electron transfer is an ideal goal of electrical wiring. It can be achieved by employing nanomaterials which suggest the direct electron transfer between enzyme active site and electrode. This wiring strategy is based on the shortening of the electronic contact of the enzyme and electrode (a short distance of ~1.5 nm) where the redox center of the enzyme can be regenerated directly by the electrode [52]. Therefore, this strategy can maximize the performance of bioelectronics. The engineering needs to consider the position of the active site inside the protecting protein and the conformation of the protein in order to wire the conducting materials with the redox center. This still remains the most challenging topic.

Several variables also affect the response nature of enzyme bioelectronics. Consideration of the fundamental theory of their functions will help to improve their performances. A key well-known model of enzyme behaviors is Michaelis-Menten kinetics, $V_0 = V_{max}\frac{[S]}{K_m + [S]}$, where V_0, V_{max}, K_m, and $[S]$ are the initial velocity of the reaction, the maximal rate of the reaction, the Michaelis-Menten constant, and the concentration of the substrate, respectively. In general, it is desirable to engineer the biointerface electrode system to obtain high V_m and low K_m (good affinity). However, dynamic range is also a crucial characteristic for wearable biosensors. Traditionally, dilution or preconcentration can be used to adjust the level of the target to be fit in the linear range of the sensor; nonetheless, manipulating such processes for on-skin applications is sophisticated. Therefore, diffusion-limiting membranes may be a useful solution to tune the dynamic range. The linear range can be extended by coating a thin membrane over an active enzyme layer since the sensor response is controlled by the analyte diffusion and not by the nonlinear characteristic of enzyme kinetics. Nevertheless, it should be noted that coating may lower the sensitivity and cause slow response time.

In addition, extra membranes can be a biocompatible barrier to address challenges from biofouling and interferents, especially when electrochemical operations are made in real matrices, samples, such as sweat. A perfluorinated sulfonated membrane (Nafion®) is an example membrane, which is also easy to drop-cast. This coating membrane can protect the enzymatic layer and also prevent anionic interferents, such as ascorbate [53].

Shelf life and operational stabilities of enzymatic electrodes are among the most critical challenges. The enzyme and active materials, such as mediators, can also leach during operations. Extensive studies have been made to improve enzyme bioelectrodes, such as by crosslinking hydrogels in the presence of the enzyme [54, 55]. Such crosslinking can entrap the enzyme to be more stable; moreover, this way enhances the loading of the enzyme, while the three-dimensional structure can facilitate the transport of analytes or biofuels, improving bioelectrode functions. Nevertheless, crosslinking enzyme or covalent binding of the enzyme can change the conformation of the enzyme and thus affect the activity [56]. Furthermore, one alternative to stabilize the enzyme electrode is the addition of stabilizers, such as polyelectrolytes, dextrans, glycerol, polyethyleneimine, and hydrophobic oils [57–59]. For instance, hydrophobic mineral oil or silicone grease can be used to minimize enzyme denaturation [58, 59]. The pasting liquid helps to lower protein mobility, maintain conformational rigidity of enzymes, and barrier to hydronium ions from acid environments. This strategy can stabilize many enzymes, such as GOx, LOx, AOx, horseradish peroxidase, amino acid oxidase, and polyphenol oxidase.

Increasing enzyme loading can also improve the performance of biocatalytic devices. Employing high surface nanomaterials is useful to enhance the surface loading of the target catalyst. A graphene-based electrode is a good example platform to offer a high enzyme loading (1.1 nmol cm^{-2}); in addition, it offers a fast heterogeneous electron transfer rate (k_s) of 2.8 s^{-1} [60]. Moreover, CNTs, which have high conductivity and specific surface, represent an outstanding candidate nanomaterial for electrochemical wiring [38, 61]. The thin nanoscale structure can intimately incorporate with the active enzyme. Adsorption of GOx on CNTs provides the apparent k_s, of 1.5 s^{-1} [62]. The k_s of GOx at the hybrid biocomposite can be as high as 11.2 s^{-1} [63]. Therefore, mediatorless bioelectrodes with excellent electron transfer could be demonstrated. Their high three-dimensional architecture also offers an enhanced loading of enzyme and/or redox mediator immobilizations. As a result, this can enlarge the current output from the biosensor or BFCs. Importantly, for BFCs, the maximized OCV and current density could be observed [43]. This BFC consists of a GOx/catalase/CNT bioanode and laccase/CNT biocathode without additional mediators. The CNT/enzyme matrix was compressed together under high hydraulic pressure (10 kN). The resulting output in an air-saturated electrolyte (200 mM glucose in 0.2 M phosphate buffer solution, pH 7 at room temperature) after 3 days displayed a high maximum OCV of 1 V. Note that the GOx/catalase/CNT bioanode and the laccase/CNT biocathode showed OCV values of −0.35 and +0.6 V, respectively.

Importantly, biofluids from the skin (such as sweat and extracted interstitial fluids) contain a variety of chemicals that can inhibit enzyme activity, reflecting challenges in biosensing and BFC functions in real-time on-body applications. For instance, heavy metals can be found in sweat as the body expels chemicals or balances the charges. One example is Cu^{2+} which has been reported as an inhibitor to deactivate the enzyme. The Cu^{2+} in sweat can be in a range of 1.6–16 μM [11]. 0.1 μM Cu^{2+} could decrease the OCV value of the glucose BFC [64]. However, this enzyme-inhibitor electrochemical behavior is analytically attractive toward the development of self-powered biosensors, such as for direct heavy metal screening or indirect cysteine monitoring. For example, cysteine prefers to bind with Cu^{2+} via the Cu-S bond; this superior conjugation between cysteine and Cu^{2+} removes metal ions from the bioanode, consequently turning on the OCV.

3.4 Effects of oxygen fluctuations on electrochemical performances

Since the O_2 level in biofluids may vary, first-generation biosensors, employing O_2-dependent mechanism, are subject to inaccuracy. This issue can be addressed by using fluorocarbon pasting liquids to supply internal O_2 [65]. Using redox mediator as a second-generation sensor is another way to eliminate this error. Furthermore, FAD-dependent glucose dehydrogenase is an option to address O_2-dependent problems due to its O_2-insensitive nature, compared with GOx [49]. In addition, because of the high rate of homogeneous electron transfer rate between GOx and oxygen, GOx prefers to transfer electrons to oxygen rather than to the electrode, causing undesirable O_2 competition effect [66]. Moreover, for BFCs and self-powered sensors, the commonly used ORR cathode may cause the error under anaerobic conditions. The use of Ag_2O/Ag redox cathode, which does not depend on ORR, can be used to operate BFCs, mitigating the possible O_2 errors [30, 67]. Note that the reduction potential of Ag_2O/Ag (0.342 V vs. SHE) is close to that of O_2/OH^- (0.401 V vs. SHE) at pH 7. Moreover, using O_2-rich cathode is another possible option to mitigate O_2-deficit effects [68].

3.5 On-skin biofluid extraction: electrical-based approaches

Each person has 2.03 million sweat glands; sweat gland densities vary broadly across the skin surface and subjects, ranging from 16 to 530 glands cm^{-2} [11, 13, 69]. Normally, during exercise, sweat can be secreted around 20 nL $gland^{-1}$ min^{-1} [11]. For example, the forehead or arm can generate sweat around 3 μL cm^{-2} or even lower. The fluctuation of sweat rate is also related to numerous factors, such as activity intensity and hydration level. Therefore, the limited volume of sweat causes a challenge in sweat analysis and operations. This leads to the development of miniaturized skin-worn electrochemical devices that can be practical in such small dead volume. For instance, the textile-based energy-harvesting BFC requires sweat volume per area of 40 μL cm^{-2} to deliver steady outputs [31]. Designing a capillary chamber is a possible route for low-volume electroanalytical systems [70].

In addition to a passive way to collect sweat, one strategy is an active electrical-based approach, called "iontophoresis" [71, 72]. This active strategy offers

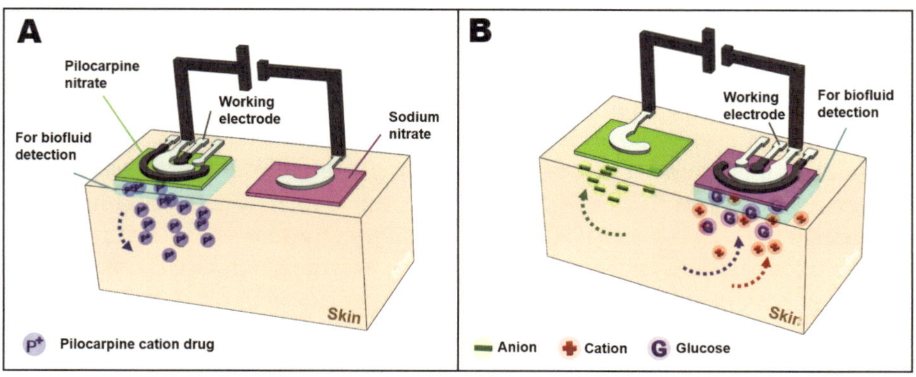

Figure 5.
Electrical-based strategies using iontophoretic electrodes to extract biofluids, including (A) pilocarpine iontophoresis and (B) reversed iontophoresis.

on-demand sweat generation as the device can be placed to a local skin target. There are two main approaches to extract sweat: (1) iontophoresis with pilocarpine drug and (2) reversed iontophoresis without the drug. These are attractive routes for continuous sweat analysis.

First, pilocarpine iontophoresis can be used to stimulate the sweat. In principle, a small electrical current is applied to enable the pilocarpine administration across the epidermis as illustrated in **Figure 5A**. For example, the tattoo-based enzymatic alcohol sensor consists of a pair of electrodes located in contact with the skin surface. Small constant current (0.2 mA cm^{-2}) was applied through the cryogel material containing pilocarpine at the anode (positive) iontophoretic side [18]. The applied electrical force will push the pilocarpine drug, which possesses a large positive charge, to eventually enter into the skin. Such transdermal drug delivery of pilocarpine can induce the local sweat, sufficient for the subsequent electrochemical detection. In addition, interstitial fluid (ISF) located under the skin can be extracted. Without this iontophoretic strategy, it is challenging to access ISF through wearable technology.

Second, the reversed iontophoresis without pilocarpine drug can be used to extract relevant analytes, such as glucose [17]. For instance, as presented in **Figure 5B**, a current (0.2 mA cm^{-2}) is applied to extract glucose in ISF. During the reverse iontophoresis process, glucose is pulled out at the negative iontophoretic compartment. Even though glucose holds no charge, the inherent permiselective characteristic of the skin prefers to transport positive species, allowing such glucose extraction. Applying electric field on mobile electric charge can cause Coulombic force, leading to a net convective flow in the skin from the anode to cathode direction. Accordingly, dissolved analytes (e.g., glucose) are also moved toward the cathode where they can be extracted and monitored. Therefore, the glucose amperometric working electrode, adjacent to the cathodic iontophoretic side, can detect the glucose level from the extracted sample.

4. Conclusions and future prospects

This chapter has reviewed some examples of new trends of skin-worn enzyme-based electrochemical systems, focusing on biosensors, BFC, and self-powered sensors. The existing systems provide significant advances toward the painless and point-of-care applications and personalized electrochemical biodevices, which was not possible without such new biodevices. However, researchers still face many challenges, such as electrochemistry, electrical wiring of enzymes, enzyme behaviors, the fabrication of stretchable electrodes, O_2 fluctuations in biofluids, interferences, and difficulty in sweat extraction. Moreover, the workability and reliability of biodevices can be limited due to the limited fluctuating and volume of biofluids. In order to avoid frequent recalibrations, the stability of biodevices or self-calibration systems are also important. Precise electrochemical functions for on-skin applications are still very challenging. Therefore, it is required careful attention to address all challenges in order to advance such wearable technologies.

Although main skin-worn BFCs have been driven by glucose and lactate fuels, it is interesting to explore new opportunities, such as from alcohol-based BFCs, where the bioanode can be functionalized with alcohol dehydrogenases. Future efforts may be made to expand the spectrum of current concepts. New integrated devices can be achieved by designing multifunctional sensors that can provide informative series of personalized data. This will require the incorporation of big-data analysis and Internet of things (IoT) to build up integrated networks and personalized baselines of each wearer. Big data collected from networks and individuals can then warn the user whether the body is in a healthy and equilibrium state or not. It is expected that developing new electrochemical biodevices will eventually track "fingerprints" of various pathologies and disorders. This aims toward wearable systems for early disease diagnosis. Moreover, full closed-loop concepts such as biocomputing logic gate, sensing, and therapeutic systems can also be further exploited in the integration of biosensors, BFCs, and drug delivery devices, in order to obtain both diagnostic and therapeutic applications. The next success of wearable biodevices needs the hybrid of multidiscipline, including physiological medicine, electronics, electrochemistry, bio- and nanoengineering, and computer science. These continued collaborative efforts will open fantastic opportunities for addressing current challenges and step further to create novel wearable devices and acquire comprehensive big data. Ultimately, it is expected that innovative wearable electrochemical technologies and new findings will contribute to revolutionizing diverse personalized wearables and biomedical applications.

Acknowledgement

The author would like to acknowledge Hassler Bueno for proof reading.

Author details

Itthipon Jeerapan[1,2]

1 Department of Chemistry, Faculty of Science, Prince of Songkla University, Hat Yai, Songkla, Thailand

2 Department of NanoEngineering, University of California San Diego, La Jolla, California, United States

*Address all correspondence to: itthipon.j@psu.ac.th

References

[1] Clark LC Jr, Lyons C. Electrode systems for continuous monitoring in cardiovascular surgery. Annals of the New York Academy of Sciences. 1962;**102**:29-45

[2] Kim J, Jeerapan I, Sempionatto JR, Barfidokht A, Mishra RK, Campbell AS, et al. Wearable bioelectronics: Enzyme-based body-worn electronic devices. Accounts of Chemical Research. 2018;**51**:2820-2828

[3] Wilson R, Turner APF. Glucose oxidase: An ideal enzyme. Biosensors & Bioelectronics. 1992;**7**:165-185

[4] Wang J. Electrochemical Glucose Biosensors. Chemical Reviews. 2008;**108**:814-825

[5] Meredith MT, Minteer SD. Biofuel cells: Enhanced enzymatic Bioelectrocatalysis. Annual Review of Analytical Chemistry. 2012;**5**:157-179

[6] Campàs M, Prieto-Simón B, Marty J-L. A review of the use of genetically engineered enzymes in electrochemical biosensors. Seminars in Cell & Developmental Biology. 2009;**20**:3-9

[7] Bandodkar AJ, Wang J. Wearable biofuel cells: A review. Electroanalysis. 2016;**28**:1188-1200

[8] Fu L, Liu J, Hu Z, Zhou M. Recent advances in the construction of biofuel cells based self-powered electrochemical biosensors: A review. Electroanalysis. 2018;**30**:2535-2550

[9] Yeknami AF, Wang X, Jeerapan I, Imani S, Nikoofard A, Wang J, et al. A 0.3-V CMOS biofuel-cell-powered wireless glucose/lactate biosensing system. IEEE Journal of Solid-State Circuits. 2018;**53**:3126-3139

[10] Ray T, Choi J, Reeder J, Lee SP, Aranyosi AJ, Ghaffari R, et al. Soft, skin-interfaced wearable systems for sports science and analytics. Current Opinion in Biomedical Engineering. 2019;**9**:47-56

[11] Bariya M, Nyein HYY, Javey A. Wearable sweat sensors. Nature Electronics. 2018;**1**:160-171

[12] Mena-Bravo A, Luque de Castro MD. Sweat: A sample with limited present applications and promising future in metabolomics. Journal of Pharmaceutical and Biomedical Analysis. 2014;**90**:139-147

[13] Sonner Z, Wilder E, Heikenfeld J, Kasting G, Beyette F, Swaile D, et al. The microfluidics of the eccrine sweat gland, including biomarker partitioning, transport, and biosensing implications. Biomicrofluidics. 2015;**9**:031301

[14] Bandodkar AJ, Jeerapan I, Wang J. Wearable chemical sensors: Present challenges and future prospects. ACS Sensors. 2016;**1**:464-482

[15] Davis PB. Cystic fibrosis since 1938. American Journal of Respiratory and Critical Care Medicine. 2006;**173**:475-482

[16] Mishra RK, Hubble LJ, Martín A, Kumar R, Barfidokht A, Kim J, et al. Wearable flexible and stretchable glove biosensor for on-site detection of organophosphorus chemical threats. ACS Sensors. 2017;**2**:553-561

[17] Bandodkar AJ, Jia W, Yardımcı C, Wang X, Ramirez J, Wang J. Tattoo-based noninvasive glucose monitoring: A proof-of-concept study. Analytical Chemistry. 2015;**87**:394-398

[18] Kim J, Jeerapan I, Imani S, Cho TN, Bandodkar A, Cinti S, et al. Noninvasive alcohol monitoring using a wearable tattoo-based Iontophoretic-biosensing system. ACS Sensors. 2016;**1**:1011-1019

[19] Kim J, Sempionatto JR, Imani S, Hartel MC, Barfidokht A, Tang G, et al. Simultaneous monitoring of sweat and interstitial fluid using a single wearable biosensor platform. Advanced Science. 2018;**5**:1800880

[20] Martín A, Kim J, Kurniawan JF, Sempionatto JR, Moreto JR, Tang G, et al. Epidermal microfluidic electrochemical detection system: Enhanced sweat sampling and metabolite detection. ACS Sensors. 2017;**2**:1860-1868

[21] Samavat S, Lloyd J, O'Dea L, Zhang W, Preedy E, Luzio S, et al. Uniform sensing layer of immiscible enzyme-mediator compounds developed via a spray aerosol mixing technique towards low cost minimally invasive microneedle continuous glucose monitoring devices. Biosensors & Bioelectronics. 2018;**118**:224-230

[22] Gowers SAN, Freeman DME, Rawson TM, Rogers ML, Wilson RC, Holmes AH, et al., Development of a minimally invasive microneedle-based sensor for continuous monitoring of β-Lactam antibiotic concentrations in vivo. ACS Sensors. 2019. DOI: 10.1021/acssensors.9b00288

[23] Gao W, Emaminejad S, Nyein HYY, Challa S, Chen K, Peck A, et al. Fully integrated wearable sensor arrays for multiplexed in situ perspiration analysis. Nature. 2016;**529**:509-514

[24] Lee H, Song C, Hong YS, Kim MS, Cho HR, Kang T, et al. Wearable/disposable sweat-based glucose monitoring device with multistage transdermal drug delivery module. Science Advances. 2017;**3**:e1601314

[25] Huang X, Zhang L, Zhang Z, Guo S, Shang H, Li Y, et al. Wearable biofuel cells based on the classification of enzyme for high power outputs and lifetimes. Biosensors & Bioelectronics. 2019;**124-125**:40-52

[26] Katz E, Willner I, Kotlyar AB. A non-compartmentalized glucose | O2 biofuel cell by bioengineered electrode surfaces. Journal of Electroanalytical Chemistry. 1999;**479**:64-68

[27] Jia W, Valdés-Ramírez G, Bandodkar AJ, Windmiller JR, Wang J. Epidermal biofuel cells: Energy harvesting from human perspiration. Angewandte Chemie, International Edition. 2013;**52**:7233-7236

[28] Berchmans S, Bandodkar AJ, Jia W, Ramírez J, Meng YS, Wang J. An epidermal alkaline rechargeable Ag–Zn printable tattoo battery for wearable electronics. Journal of Materials Chemistry A. 2014;**2**:15788-15795

[29] Bandodkar AJ, Jeerapan I, You J-M, Nuñez-Flores R, Wang J. Highly stretchable fully-printed CNT-based electrochemical sensors and biofuel cells: Combining intrinsic and design-induced Stretchability. Nano Letters. 2016;**16**:721-727

[30] Jeerapan I, Sempionatto JR, Pavinatto A, You J-M, Wang J. Stretchable biofuel cells as wearable textile-based self-powered sensors. Journal of Materials Chemistry A. 2016;**4**:18342-18353

[31] Lv J, Jeerapan I, Tehrani F, Yin L, Silva-Lopez CA, Jang J-H, et al. Sweat-based wearable energy harvesting-storage hybrid textile devices. Energy & Environmental Science. 2018;**11**:3431-3442

[32] Yu Y, Zhai J, Xia Y, Dong S. Single wearable sensing energy device based on photoelectric biofuel cells for simultaneous analysis of perspiration and illuminance. Nanoscale. 2017;**9**:11846-11850

[33] Ogawa Y, Kato K, Miyake T, Nagamine K, Ofuji T, Yoshino S, et al. Organic transdermal Iontophoresis patch with built-in biofuel cell.

Advanced Healthcare Materials. 2015;**4**:506-510

[34] Bandodkar AJ, You J-M, Kim N-H, Gu Y, Kumar R, Mohan AMV, et al. Soft, stretchable, high power density electronic skin-based biofuel cells for scavenging energy from human sweat. Energy & Environmental Science. 2017;**10**:1581-1589

[35] Liu Y, Pharr M, Salvatore GA. Lab-on-skin: A review of flexible and stretchable electronics for wearable health monitoring. ACS Nano. 2017;**11**:9614-9635

[36] Pailler-Mattei C, Bec S, Zahouani H. In vivo measurements of the elastic mechanical properties of human skin by indentation tests. Medical Engineering & Physics. 2008;**30**:599-606

[37] Amjadi M, Kyung K-U, Park I, Sitti M. Stretchable, skin-mountable, and wearable strain sensors and their potential applications: A review. Advanced Functional Materials. 2016;**26**:1678-1698

[38] Wang J. Carbon-nanotube based electrochemical biosensors: A review. Electroanalysis. 2005;**17**:7-14

[39] Laoui T. Mechanical and thermal properties of styrene butadiene rubber—Functionalized carbon nanotubes nanocomposites. Fullerenes, Nanotubes, and Carbon Nanostructures. 2013;**21**:89-101

[40] Abellán-Llobregat A, Jeerapan I, Bandodkar A, Vidal L, Canals A, Wang J, et al. A stretchable and screen-printed electrochemical sensor for glucose determination in human perspiration. Biosensors & Bioelectronics. 2017;**91**:885-891

[41] Ahmadi MM, Jullien GA. A very low power CMOS potentiostat for bioimplantable applications. In: Fifth International Workshop on System-on-Chip for Real-Time Applications (IWSOC'05); 2005; pp. 184-189

[42] Cass AEG, Davis G, Francis GD, Hill HAO, Aston WJ, Higgins IJ, et al. Ferrocene-mediated enzyme electrode for amperometric determination of glucose. Analytical Chemistry. 1984;**56**:667-671

[43] Agnès C, Holzinger M, Le Goff A, Reuillard B, Elouarzaki K, Tingry S, et al. Supercapacitor/biofuel cell hybrids based on wired enzymes on carbon nanotube matrices: Autonomous reloading after high power pulses in neutral buffered glucose solutions. Energy & Environmental Science. 2014;**7**:1884-1888

[44] Sokic-Lazic D, de Andrade AR, Minteer SD. Utilization of enzyme cascades for complete oxidation of lactate in an enzymatic biofuel cell. Electrochimica Acta. 2011;**56**:10772-10775

[45] Ferri S, Kojima K, Sode K. Review of glucose oxidases and glucose dehydrogenases: A Bird's eye view of glucose sensing enzymes. Journal of Diabetes Science and Technology. 2011;**5**:1068-1076

[46] Le Goff A, Holzinger M. Molecular engineering of the bio/nano-interface for enzymatic electrocatalysis in fuel cells. Sustainable Energy & Fuels. 2018;**2**:2555-2566

[47] Saboe PO, Conte E, Farell M, Bazan GC, Kumar M. Biomimetic and bioinspired approaches for wiring enzymes to electrode interfaces. Energy & Environmental Science. 2017;**10**:14-42

[48] Milton RD, Lim K, Hickey DP, Minteer SD. Employing FAD-dependent glucose dehydrogenase within a glucose/oxygen enzymatic fuel cell operating

in human serum. Bioelectrochemistry. 2015;**106**:56-63

[49] Tsujimura S, Kojima S, Kano K, Ikeda T, Sato M, Sanada H, et al. Novel FAD-dependent glucose dehydrogenase for a dioxygen-insensitive glucose biosensor. Bioscience, Biotechnology, and Biochemistry. 2006;**70**:654-659

[50] Yoshino S, Miyake T, Yamada T, Hata K, Nishizawa M. Molecularly ordered bioelectrocatalytic composite inside a film of aligned carbon nanotubes. Advanced Energy Materials. 2013;**3**:60-64

[51] Jeerapan I, Ciui B, Martin I, Cristea C, Sandulescu R, Wang J. Fully edible biofuel cells. Journal of Materials Chemistry B. 2018;**6**:3571-3578

[52] de Poulpiquet A, Ciaccafava A, Lojou E. New trends in enzyme immobilization at nanostructured interfaces for efficient electrocatalysis in biofuel cells. Electrochimica Acta. 2014;**126**:104-114

[53] Cordeiro CA, de Vries MG, Cremers TIFH, Westerink BHC. The role of surface availability in membrane-induced selectivity for amperometric enzyme-based biosensors. Sensors and Actuators B: Chemical. 2016;**223**:679-688

[54] Zhou J, Liao C, Zhang L, Wang Q, Tian Y. Molecular hydrogel-stabilized enzyme with facilitated electron transfer for determination of H_2O_2 released from live cells. Analytical Chemistry. 2014;**86**:4395-4401

[55] Chakraborty D, McClellan E, Hasselbeck R, Barton SC. Characterization of enzyme-redox hydrogel thin-film electrodes for improved utilization. Journal of the Electrochemical Society. 2014;**161**:H3076-H3082

[56] Secundo F. Conformational changes of enzymes upon immobilisation. Chemical Society Reviews. 2013;**42**:6250-6261

[57] Rocchitta G, Spanu A, Babudieri S, Latte G, Madeddu G, Galleri G, et al. Enzyme biosensors for biomedical applications: Strategies for safeguarding analytical performances in biological fluids. Sensors. 2016;**16**:780

[58] Wang J, Liu J, Cepra G. Thermal stabilization of enzymes immobilized within carbon paste electrodes. Analytical Chemistry. 1997;**69**:3124-3127

[59] Wang J, Musameh M, Mo J-W. Acid stability of carbon paste enzyme electrodes. Analytical Chemistry. 2006;**78**:7044-7047

[60] Kang X, Wang J, Wu H, Aksay IA, Liu J, Lin Y. Glucose oxidase–graphene–chitosan modified electrode for direct electrochemistry and glucose sensing. Biosensors & Bioelectronics. 2009;**25**:901-905

[61] Holzinger M, Le Goff A, Cosnier S. Carbon nanotube/enzyme biofuel cells. Electrochimica Acta. 2012;**82**:179-190

[62] Cai C, Chen J. Direct electron transfer of glucose oxidase promoted by carbon nanotubes. Analytical Biochemistry. 2004;**332**:75-83

[63] Palanisamy S, Cheemalapati S, Chen S-M. Amperometric glucose biosensor based on glucose oxidase dispersed in multiwalled carbon nanotubes/graphene oxide hybrid biocomposite. Materials Science and Engineering: C. 2014;**34**:207-213

[64] Hou C, Fan S, Lang Q, Liu A. Biofuel cell based self-powered sensing platform for l-cysteine detection. Analytical Chemistry. 2015;**87**:3382-3387

[65] Wang J, Lu F. Oxygen-rich oxidase enzyme electrodes for operation

in oxygen-free solutions. Journal of the American Chemical Society. 1998;**120**:1048-1050

[66] Filip J, Tkac J. Is graphene worth using in biofuel cells? Electrochimica Acta. 2014;**136**:340-354

[67] Yu Y, Xu M, Bai L, Han L, Dong S. Recoverable hybrid enzymatic biofuel cell with molecular oxygen-independence. Biosensors & Bioelectronics. 2016;**75**:23-27

[68] Jeerapan I, Sempionatto JR, You J-M, Wang J. Enzymatic glucose/ oxygen biofuel cells: Use of oxygen-rich cathodes for operation under severe oxygen-deficit conditions. Biosensors & Bioelectronics. 2018;**122**:284-289

[69] Taylor NA, Machado-Moreira CA. Regional variations in transepidermal water loss, eccrine sweat gland density, sweat secretion rates and electrolyte composition in resting and exercising humans. Extreme Physiology & Medicine. 2013;**2**:4

[70] Cai X, Klauke N, Glidle A, Cobbold P, Smith GL, Cooper JM. Ultra-low-volume, real-time measurements of lactate from the single heart cell using microsystems technology. Analytical Chemistry. 2002;**74**:908-914

[71] Choi D-H, Thaxton A, Jeong Ic, Kim K, Sosnay PR, Cutting GR, et al. Sweat test for cystic fibrosis: Wearable sweat sensor vs. standard laboratory test. Journal of Cystic Fibrosis. 2018;**17**:e35-e38

[72] Choi J, Ghaffari R, Baker LB, Rogers JA. Skin-interfaced systems for sweat collection and analytics. Science Advances. 2018;**4**:eaar3921

Chapter 4

Breathing Monitoring and Pattern Recognition with Wearable Sensors

Taisa Daiana da Costa, Maria de Fatima Fernandes Vara, Camila Santos Cristino, Tyene Zoraski Zanella, Guilherme Nunes Nogueira Neto and Percy Nohama

Abstract

This chapter introduces the anatomy and physiology of the respiratory system, and the reasons for measuring breathing events, particularly, using wearable sensors. Respiratory monitoring is vital including detection of sleep apnea and measurement of respiratory rate. The automatic detection of breathing patterns is equally important in other respiratory rehabilitation therapies, for example, magnetic resonance exams for respiratory triggered imaging, and synchronized functional electrical stimulation. In this context, the goal of many research groups is to create wearable devices able to monitor breathing activity continuously, under natural physiological conditions in different environments. Therefore, wearable sensors that have been used recently as well as the main signal processing methods for breathing analysis are discussed. The following sensor technologies are presented: acoustic, resistive, inductive, humidity, acceleration, pressure, electromyography, impedance, and infrared. New technologies open the door to future methods of noninvasive breathing analysis using wearable sensors associated with machine learning techniques for pattern detection.

Keywords: breathing analysis, sensors, wearable device, respiration monitoring, pattern recognition

1. Introduction

Wearable devices mean whatever a person can wear since they do not restrict daily activities or mobility [1]. Recently, progress has been made in the use of wearable sensors for breathing monitoring devices, so that it is considered a promising area [2]. Many applications, including sleep monitoring [3], breathing pattern detection, and respiratory rate detection [4, 5], require comfortable and wearable devices that patients can wear in their homes, if possible, for continuous monitoring and storage of relevant data. Other requirements for wearable devices involve (i) the ability to share patient data with healthcare professionals, researchers, and family, (ii) very low energy consumption and long battery autonomy, and (iii) wireless communication with other devices [1, 6].

The main topics for the development of wearable devices for breathing monitoring and pattern detection are discussed in this chapter.

1.1 Why is it important to monitor breathing activity with wearable devices?

The development of wearable devices to monitor breathing activity allows giving rise to various medical care services. For example, considering people with asthma or chronic obstructive pulmonary disease, the environmental conditions directly affect their breathing, and a wearable device is able to continually measure air quality and pulmonary function [7]. The device could trigger alarm functions for drug uptake, contact a general practitioner for an appointment, or call emergency services [8].

The measurement of air quality is important, as pollutant exposure can lead to acute asthma attacks [7]. This happens usually after days under exposure. If a system detects pollutant exposure, it can warn the person and help to prevent attacks [7, 9].

Other applications of wearable devices include sleep monitoring for apnea detection [3], speaking detection as an indicator of social interaction [10], respiratory impedance [8], etc. The detection and tracking of respiratory movement for image-guided chest and abdomen radiotherapy, for compensation of movement during treatment, are additional uses of wearable devices [11]. Moreover, researchers have studied ways to develop smart fabrics, which are comfortable and nonintrusive, for different applications such as healthcare, sports, and military scenarios [5].

1.2 What is important to know for the development of a wearable device for breathing monitoring and pattern detection?

The creation of these wearable devices requires understanding the anatomy and physiology of the respiratory system. The knowledge about its structure and function leads to the development of devices that do not interfere with respiratory mechanics or daily life activities. It also allows selecting the best sensors in each case. Therefore, it is important to have an overview of the main types of electronic sensors used in recent years and how they have been applied, as well as signal processing and machine learning methods.

This chapter covers these topics concisely as a guide for people interested in developing wearable devices for respiratory monitoring. The next section introduces the anatomy and physiology of the respiratory system. The sections 3, 4, and 5 discuss, respectively, the electronic sensors, signal processing methods, and machine learning techniques applied to respiratory signals for pattern recognition.

2. Anatomy and physiology: mechanics of respiration

When one thinks of breathing, the airways and the airflow come to mind. Therefore, an understanding starting with the structures involved in this process is very important.

2.1 Respiratory system

The respiratory system consists of the following structures [12, 13] (**Figure 1**):

- Nose: nasal fossae; nasal cavity; pharynx (muscle tube); larynx (cartilage tube); trachea—bifurcates into two primary bronchi, which enter the pulmonary lobes, then subdivided into progressively smaller structures: bronchioles, ducts, and alveoli (where gas exchange occurs).

- Airways: space from the nose to the bronchioles (where no gas exchange occurs). The structures up to the trachea are responsible for conducting, filtering, heating, and humidifying the air.

- Lungs: the principal organs of the respiratory system, surrounded by a membrane of connective-elastic tissue called visceral pleura. There are also the parietal pleura, which cover the thoracic cavity. Between them, there is pleural fluid, which contributes to respiratory mechanics.

Not only structures play an important role in respiration. Airflow direction delimits the breathing phases. Breathing comprises two steps. The first is the transport of oxygen (O_2) through inhalation, from the environment to the cells. The second is the transport of carbon dioxide (CO_2) from the intracellular to the environment. Breathing aims to supply the cells with adequate amounts of O_2 and withdraw CO_2 from the body to maintain homeostasis [13].

The lungs are positioned in an airtight space, and the oscillation of their pressure volume is the basis for respiratory control. The intrathoracic pressure is negative compared to the lung pressure. The lung functions as an elastic structure that resists deformation. The ability of the lung to expand is called compliance [14] and is expressed as Eq. (1).

$$C = dV/dP \tag{1}$$

Compliance requires a respiratory effort under conditions of normality. When compliance is reduced, more effort is demanded from the respiratory system, and, in more severe cases, it may lead to respiratory insufficiency.

Thorax compliance (C_T), lung compliance (C_L), and lung-thorax system compliance (C_{LT}) may be expressed by Eqs. (2), (3) and (4), respectively, according to [14].

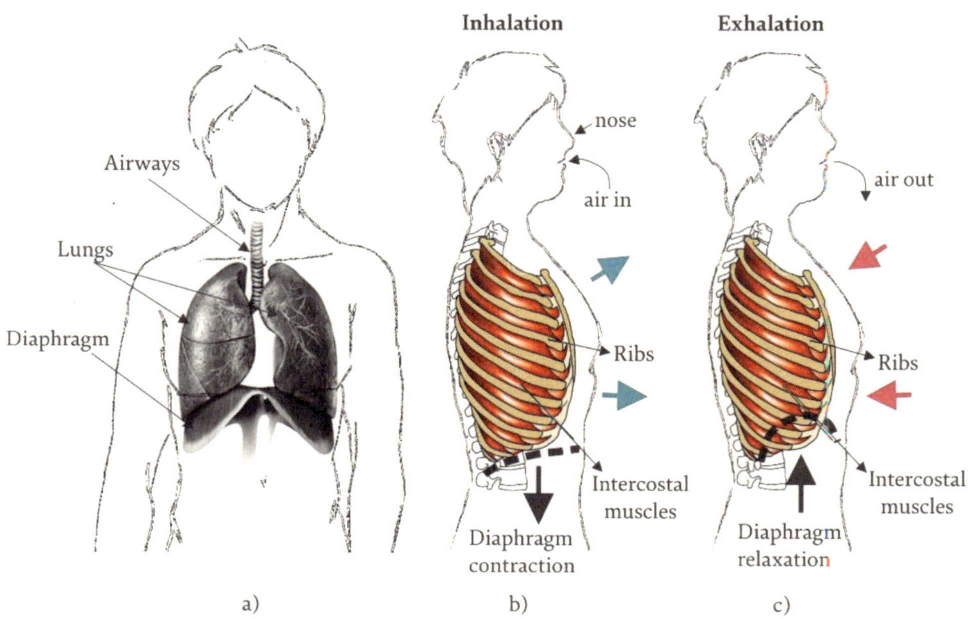

Figure 1.
Breathing process: (a) structures involved in the breathing process; (b) inhalation event; and (c) exhalation event.

$$C_T = \frac{dV}{dP_T} \tag{2}$$

$$C_L = \frac{dV}{dP_L} \tag{3}$$

$$C_{LT} = \frac{dV}{dP_{LT}} \tag{4}$$

Breathing also involves air diffusion, exchange from a more concentrated to a less concentrated medium. Poiseuille's law governs the flow resistance as expressed by Eq. (5).

$$R = \frac{8\eta L}{\pi r^4} \tag{5}$$

Where R is the flow resistance, L is the length, η is the viscosity of air, and r is the radius of the tubes.

Figure 1 shows the main structures and processes involved in breathing.

2.2 Muscles involved in breathing and their functions

The diaphragm is the most important muscle of inspiration. When it contracts, there is a decrease in intrapleural pressure and an increase in lung volume [13]. Simultaneously, an increase in abdominal pressure is transmitted to the chest through the apposition zone to expand the lower thoracic cavity. When the diaphragm contracts, the lower rib cage expands. One may observe the bucket handle movement that causes an increase in thorax transverse diameter due to the elevation of the ribs during inspiration [15]. Elevation and sternum forward movement during inspiration causes the increase of thorax anteroposterior diameter. Diaphragm contraction also contributes to increasing the longitudinal thorax diameter [12].

Scalene muscle, sternocleidomastoid muscle, and intercostal muscle are inspiration auxiliary muscles. During forced expiration, the abdominal muscles contract, and the diaphragm is pushed upward, thus causing a decrease in chest diameters. Abdominal muscle is also important for coughing [16].

2.3 Different etiologies, types, and characteristics of pathological respiratory patterns

If structural and/or functional changes occur, then adequate air transport to and from the lungs can be compromised. There are different etiologies, types, and pathological respiratory patterns in which wearable systems may assist in the characterization of movement patterns [1]. This capacity helps in the analysis of the health condition of patients, providing important additional information.

Thoracic mobility is related to the integrity of the nerve pathways and respiratory muscles [13]. In clinical practice, thoracic and abdominal amplitude measurements during respiratory movement may provide information on changes in the respiratory system or eventual diseases [17]. Some paradoxical movements may occur when patients present weakness, muscle paralysis, or chronic obstructive pulmonary disease (COPD), with pulmonary hyperinflation, among other commitments [18]. Another example is Cheyne-Stokes breathing, which is a type of central sleep apnea with an unstable breathing pattern throughout the night. It can cause changes in respiratory frequency and depth of patients with congestive heart failure [19].

Other impairments may cause changes in the thoracic and abdominal mobility relation such as dyspnea, orthopnea, alternate breathing, forced expiration, etc. Wearable systems capable of monitoring the contribution of different muscles and changes in mobility patterns can help monitor the evolution of the respiratory functional condition of a person.

2.4 Pulmonary auscultation: sounds in healthy and diseased lungs

Lung sounds occur because of air turbulence in the larger airways [15, 20]. They are the results of pulmonary vibrations and the respective airways transmitted to the thoracic wall. Sounds that occur during natural breathing differ depending on where they are acquired as well as the moment of the ventilatory cycle [20]. So, controlling where to place wearable devices and their sampling frequency and duration allows obtaining significant data from lung sounds.

Normal pulmonary sounds are classified into:

- Tracheal sound: it is audible in the region of the trachea from cervical to sternal height, having an intense and tubular sound. Inspiration is slightly shorter than expiration, with a pause between events [21].

- Bronchial sound: it is audible in the region of the bronchi, at the height of the sternal manubrium, having less intensity than the tracheal sound. The duration of inspiration and expiration is similar, with a pause between events [22].

- Bronchovesicular sound: it is audible in the first and second intercostal spaces and between the scapulae. The duration of inspiration and expiration is similar, with no pause between events [22].

- Vesicular murmur: it is audible in the peripheral regions of the lungs, having less intensity than the bronchial sound. Inspiration is longer than expiration, with no pause between events [21].

The anatomical structures may influence the sound heard during normal breathing [21].

Pathological changes in the lungs directly affect the perception of lung sounds from the airways to the thoracic surface. Abnormal lung sounds, also called adventitious noises, are classified into:

- Wheezing: it occurs with the oscillations of the bronchial pathways [22].

- Rhonchus: similar to snoring, it can be heard during inspiration and/or expiration [21].

- Crackles: they are discontinuous sounds, presented in a short and explosive manner, usually classified considering their duration and loudness, during the respiratory cycle [22].

There are other sounds and more details about each of them, and wearable systems contribute to distinguishing the different sounds in clinical practice.

The concepts presented in this section are very important for understanding the respiratory system in healthy and unhealthy conditions. Depending on the event one aims to observe, this information helps to identify the best location for sensor placement. It also contributes to a better interpretation of the respiratory signals obtained.

After this brief overview of the main concepts involving respiratory anatomy and physiology, the next section explains how wearable devices for respiratory monitoring have been made.

3. Respiratory wearable sensors

Wearable sensors for respiratory monitoring employ various types of electronic sensors that can be mounted into clothes [23], attached to belts [5, 24], fixed on the skin [3, 7], etc. There are many ways to make wearable devices and some of them are described separately by the type of primary sensor in the following sections.

3.1 Pressure sensors

We can take advantage of the events of diaphragm contraction (as shown in **Figure 1b**) and relaxation (as illustrated in **Figure 1c**) to create wearable devices based on pressure sensors. As an example, researchers have used an electromechanical film (EMFit) to develop a respiratory rate sensor designed as a belt [24] (as shown in **Figure 2a**). They attached the sensor to the belt so that the expansion of the chest during breathing applies a force to the sensor, and produces a voltage change proportional to this movement. EMfit is a capacitive pressure sensor that has a thin porous polypropylene film structure with a sensitivity of 30–170 pC/N.

Another way to use pressure sensors is to use them directly in contact with the inhaled and exhaled air pressure during breathing. The facemask introduced in [8] measures the respiratory impedance and was targeted to home and clinical applications. The solution consists of two pressure transducers, two low power consumption fans, a field-programmable gate array, and a real-time processing engine. The device is based on the forced oscillation technique (FOT), which is a nonstandardized lung function test. The idea is to use fans to input a periodic sinusoidal air pressure signal and measure the opposite force produced by the respiratory tract. With these data, respiratory resistance and compliance, as shown in Eq. (1), can be calculated and sent via Bluetooth to a smartphone (**Figure 2b**).

The EMFit sensor is less intrusive and performed well in the detection of respiratory rate. However, body movements affect the accuracy of the measurement, so the sensor only worked well for still or moderate moving patients [24]. The facemask sensor also performed well and estimated the respiratory impedance satisfactorily. Nevertheless, it was a prototype and its use was not comfortable [8].

3.2 Acoustic sensors

As seen in Section 2.4, it is possible to monitor lung sounds using acoustic sensors. Acoustic signals related to breathing are usually obtained with the sensors located close to the nose, mouth, throat, and suprasternal notch [3, 25, 26]. **Figure 3a** shows a wireless microphone that is a portable, cheap, and easy-to-use wearable device positioned next to the nose [3]. The purpose was to measure the respiratory rate in sleep. The microphone is fixed near the nose with a tape, and the signals are sent to a smartphone via wireless communication.

BodyScope was developed to record the sounds produced by the throat region in order to classify them into the following categories [25]: eating, drinking, speaking, laughing, and coughing. The developers modified a wireless headset attaching a microphone and a stethoscope chestpiece to minimize external source audio, as illustrated in **Figure 3b**. The position selected to place the sensor was close to the

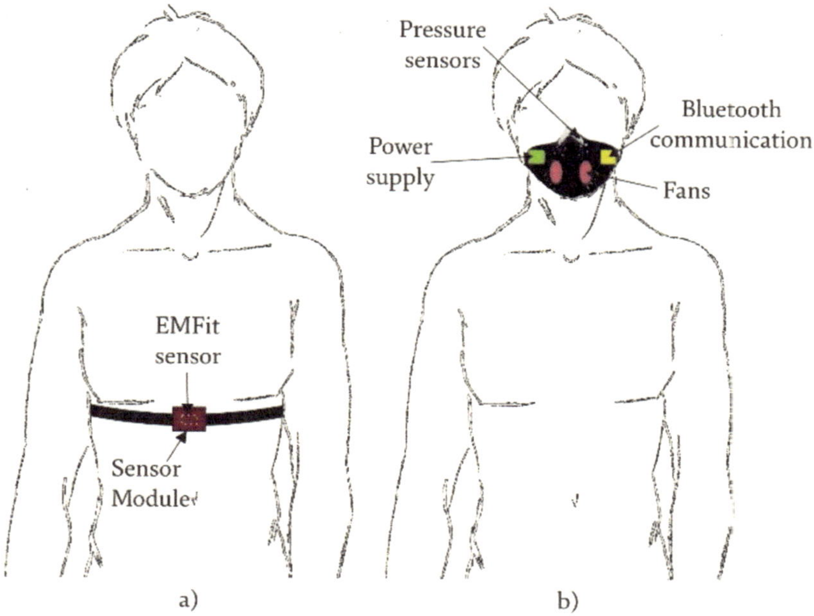

Figure 2.
Wearable pressure sensors: (a) pressure sensor (EMFit) attached to the belt and against the skin: the variations of ribcage volume during respiration compresses the sensor, producing a proportional charge [24]; (b) system developed by [8] for respiratory impedance measurement based on the forced oscillation technique.

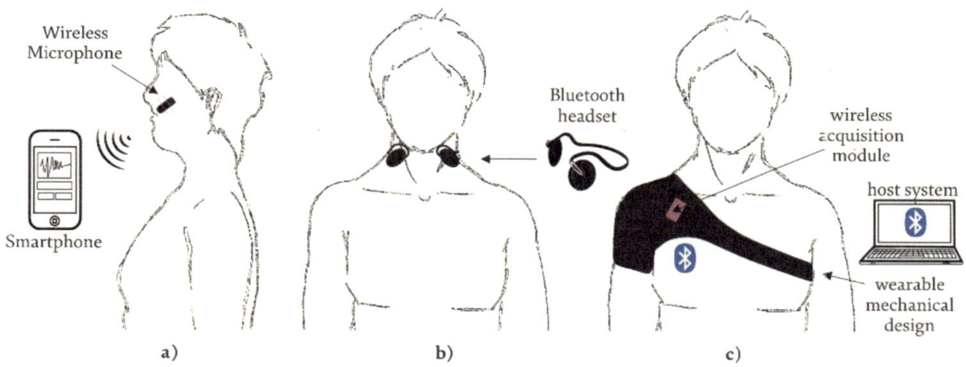

Figure 3.
Acoustic devices for respiratory monitoring: (a) a wireless microphone connected to a smartphone application [3]; (b) BodyScope system: Bluetooth headset attached with a microphone and a stethoscope chestpiece [25]; (c) a wireless acquisition module embedded into a wearable mechanical design [23] and placed over the right chest.

carotid artery region as indicated the preliminary test results. The device sends the audio signals to a computer or smartphone likewise solution shown in **Figure 3a** [3].

Figure 3c shows a real-time wheeze detector that consists of a wireless sound acquisition module, a wearable mechanical design and a host system [23]. The sensor module was an omnidirectional condenser microphone and a stethoscope bell.

A commercial repository of normal and abnormal lung sounds (referred to as the R.A.L.E lung repository) was used to implement and evaluate a wearable sensor that monitors lung sounds continuously for asthma attack detection [27]. The sensor is a microphone array for pre-filtered acoustic signal acquisition. It is an acoustic resonator array consisting of 13 paddle-shaped piezoelectric cantilevers. The results showed that accessing a repository to test for event detections did not hinder its application as a wearable system.

Acoustic wearable sensors can be very practical. However, some challenges are faced during the project design phase such as determining the optimal sensor position, canceling the acoustic ambient noise and the detection of movement artifacts. Depending on the setting, its use is not possible.

3.3 Humidity sensors

Wearable humidity sensors based on the porous graphene network (a chemical structure capable of detecting moisture) have been tested for breathing analysis [4]. The sensors are capable of sensing the human respiration, apnea, speaking, and whistle rhythm. The sensors are attached to the body with a facemask, as shown in **Figure 4**. The disadvantage of using this sensor is that long time use is also uncomfortable. It still needs some improvements to further commercialization.

3.4 Oximetry sensors

Oximetry is the technique used to measure oxygen saturation. It consists, basically, of a small infrared emitter that illuminates a small portion of the skin and a receiver that measures the light absorption depending on the oxygenated and deoxygenated blood levels [28]. Wearable oximetry sensors can be worn on the wrist, finger, head, earphones, earlobe, thigh, and ankle, and they have been widely commercialized [1] (**Figure 5**).

3.5 Acceleration sensors

Accelerometers can be used to capture the respiratory movements during inhalation and exhalation events [29]. An adhesive sensor (called BiostampRC®) made of a triaxial accelerometer that can be placed on the chest wall (**Figure 6b**) has been used [29].

Researchers adapted the EMFit-based sensor to evaluate MEMS (microelectro mechanical system) high-resolution capacitive accelerometers for the detection of respiratory rate at the same time [24]. They attached two monoaxial accelerometers to the belt as shown in **Figure 6a**.

A better signal can be obtained depending on the location of the sensor [30, 31], because people may have disorders that affect muscle contraction during breathing [32], as seen in Section 2.3. Accelerometers have found application in many areas, recently, since sensors operate in a wide spectral range and have small dimensions

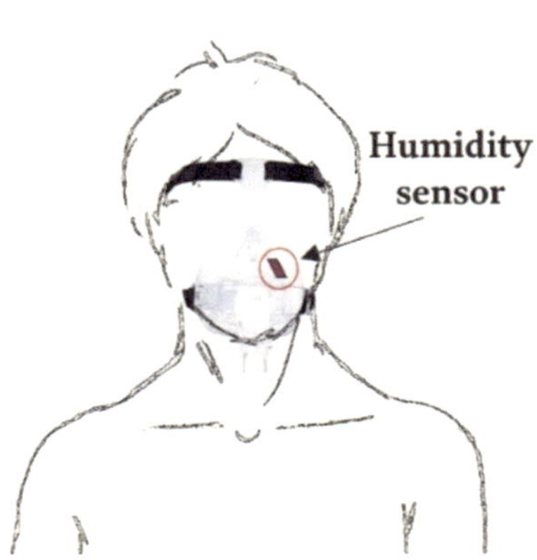

Figure 4.
Humidity sensor attached to a facemask [4].

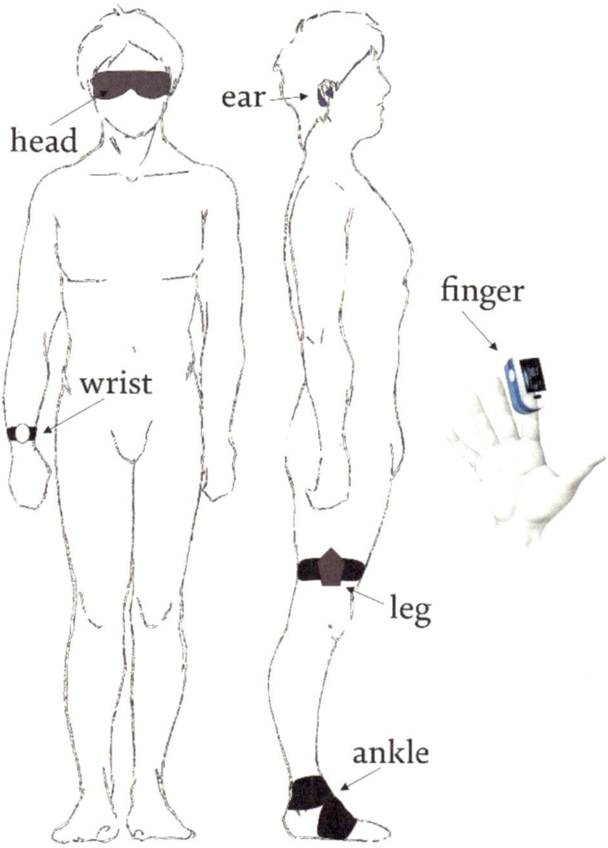

Figure 5.
Location of some oximetry wearable devices [1].

[33, 34]. In spite of that, in the clinical setting, body movement seriously influences them [35]. The sensitivity can be set to measure vibrations with amplitude varying from gross body movements to small artery pulsation [36]. Therefore, likewise applications with acoustic sensors, unwanted artifacts have to be detected in order to prevent taking decisions based on contaminated lung signals [37]. The activation of synchronized functional electrical stimulation should consider these undesirable artifacts.

3.6 Resistive sensors

Another work used a textile sensor to detect talk events based on changes in breathing patterns [10]. The solution consisted of resistive stretch sensors that are made with a conductive material and a polymer mixture. These components were attached to three different belts: upper chest, lower chest, and abdomen as illustrated in **Figure 7a**. The events of thoracic or abdominal expansion and relaxation result in variation in the resistance of the stretch sensor with this sensor configuration. The idea is that the sensor can be directly integrated into the clothing in the future.

Piezoresistive sensors can also be used for the production of wearable devices. **Figure 7b** shows an example in which a smart textile fabric for respiratory rate monitoring was developed using a conductive piezoresistivity-based yarn garment [5].

Movement artifacts are also a problem for this kind of sensor. Researchers are working on improvements to incorporate these sensors in clothes and allow for activities such as running and cycling in the future [5, 38, 39].

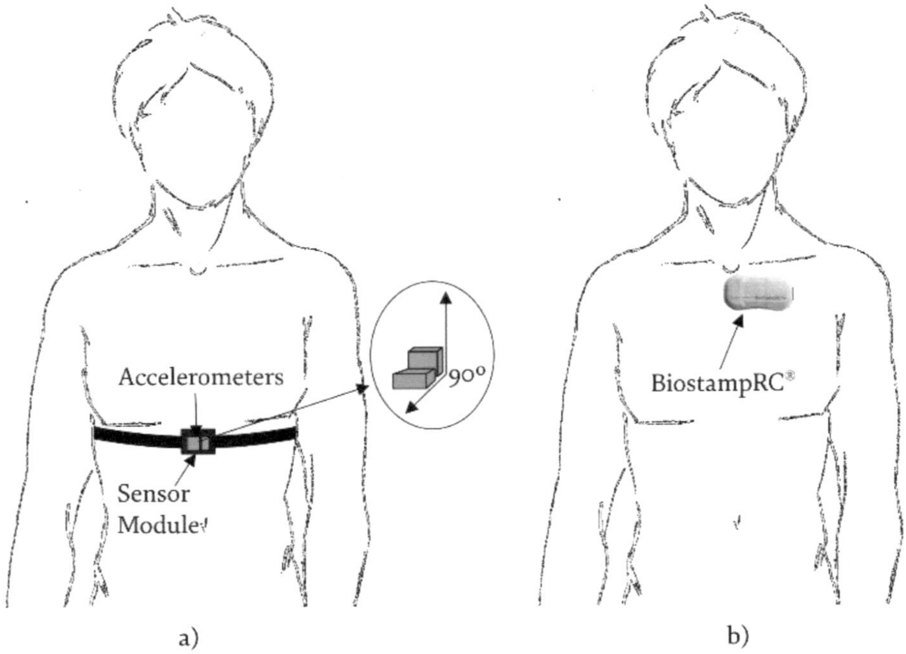

Figure 6.
(a) The 1-axis accelerometers were mounted perpendicularly and parallel relative to the chest plane [24]; (b) the BiostampRC® system.

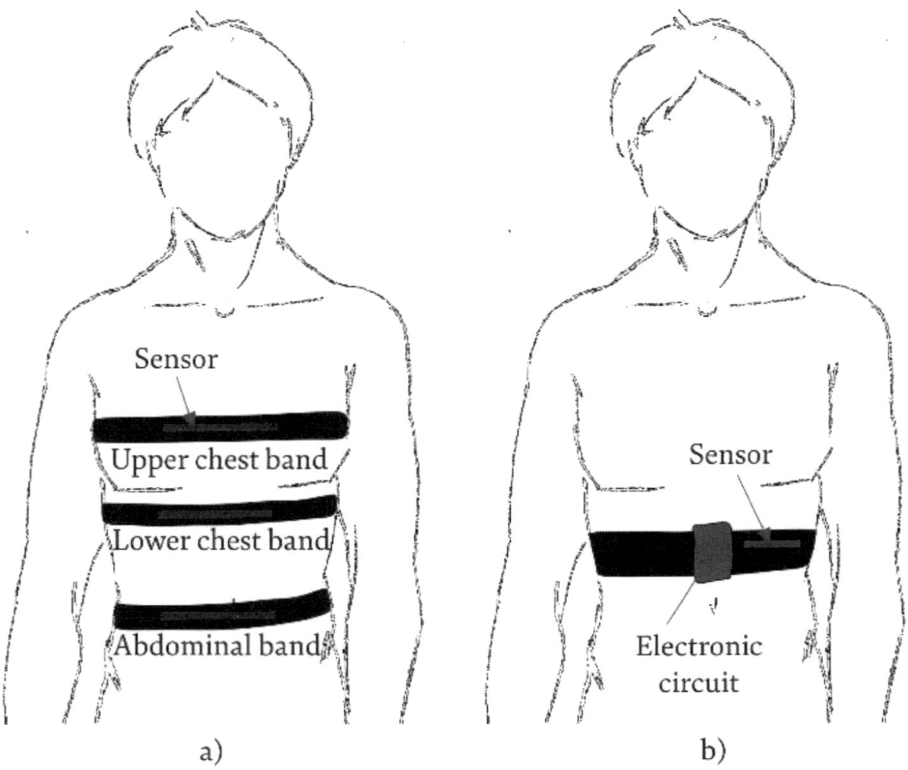

Figure 7.
(a) System consisting of different belts to monitor chest and/or abdominal breathing [10]; (b) piezoresistive sensor [5].

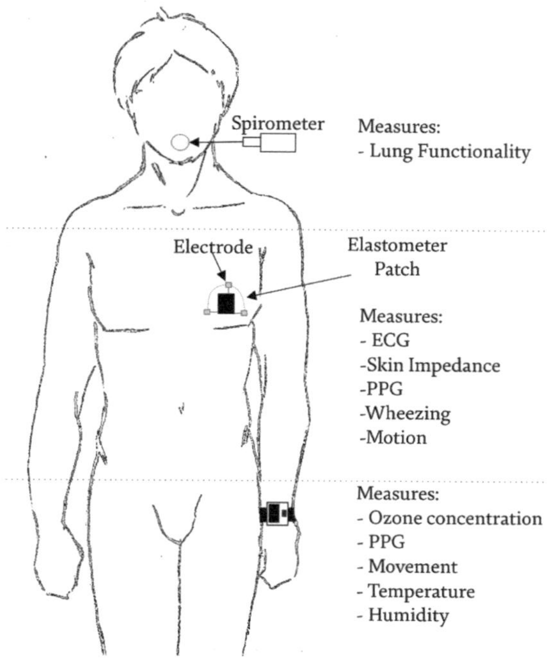

Figure 8.
Multimodal system [7].

3.7 Multimodal sensing platforms

Low-power multimodal wearable systems for the continuous monitoring of respiratory activity have been developed. **Figure 8** shows a system with a sensing platform that consists of a chest-patch, a wristband, and a handheld spirometer [7]. Its aim is to monitor health and the environment for asthma management. The chest-patch measures electrocardiogram (ECG), skin impedance, photoplethysmography (PPG), movement, and acoustic signals. The spirometer can measure forced expiratory volume in 1 s (FEV1), peak expiratory flow (PEF), and forced expiratory capacity (FVC). The wristband sensors are intended to measure ozone exposure, ambient temperature, relative humidity, PPG, and movement. The idea is to create a system for continuous long-term monitoring of the state of health and the environmental factors relevant to respiratory problems such as asthma.

This brief overview revealed that different sensors can monitor the same respiratory event and there are different ways to apply them. The sensors discussed are not limited to the applications mentioned in this chapter; they can be used in many other applications and combinations. One of the most difficult tasks is to develop a respiratory wearable device that is low cost, low power consuming, and immune to movement artifacts other than the pulmonary ones.

4. Signal processing methods for respiratory signals

4.1 Amplification

Some sensor signals have very low amplitude and need to be processed. The sensitivity of the EMFit, for example, is about 2.2–7 mV/mmHg. For signals so small, high-impedance voltage amplifiers must be used [24].

4.2 Filtering

Depending on the signal, filtering is advantageous for processing [40]. Filters are quite common in biomedical engineering applications to emphasize the spectral contents of electrophysiological signals [41]. There are signals with a well-known spectrum that researchers have extensively investigated. Once the frequency range of the signal is determined, an electronic circuit prevents unwanted energy from contributing to the processing and decision-making [42]. As an example, if the acoustic signal band frequency of interest of a solution is between 500 and 900 Hz, then a band-pass filter encompassing this spectrum is inserted into the circuit [43]. For each sensor, one filter should be placed.

Filters can be applied to minimize high-frequency noise, preserving the shape of the respiratory signal [29]. A band-pass filter with cutoff frequencies of 0.1 and 1.5 Hz was applied to compensate for possible drifts and to reduce the total noise level in the signals [10]. **Table 1** shows some types of filters used by the researchers in this area.

4.3 Analog to digital processing

Despite the advances in digital technologies, we still live in a world full of analog phenomena and human physiology is no exception. Almost all electronic biomedical devices use some kind of quantity conversion, from analog to digital. The exceptions are those devices that work entirely in analog mode.

Key factors of analog to digital conversion need to be considered in order to understand the operation of mobile wearable devices. One factor relates to Claude Shannon's [46] and Harry Nyquist's theories [47]. The sampling theory helps to determine the acquisition frequency (or sampling frequency f_s) of analog signals. To digitize a pure sinusoidal wave properly, an acquisition frequency of at least twice the maximum frequency of the analog signal must be used. Knowledge of the spectral range is therefore crucial for determining f_s.

Human electrophysiological signals are not purely sinusoidal so that the developers of biomedical systems should be far more conservative in determining f_s. Knowing the maximum frequency of the bandwidth (f_{max}) is useful, because the theory indicates to set f_s at least twice that value ($f_s \geq 2 \times f_{max}$). In some cases, however, f_s must be high enough to keep the signal's significant energy depending on the frequencies of interest. Acoustic sounds, for instance, revealed that signal power was mainly distributed below 5000 Hz [25]. The

Reference	Sensor	Type of filter	Cutoff frequencies (Hz)
[24]	Pressure and accelerometer	2nd order Butterworth low-pass filter	0.2
[44]	Acoustic	2nd order Butterworth high-pass and low-pass filters in series	20–2000
[23]	Acoustic	Band-pass filter	150–1000
[10]	Resistive	Band-pass filter	0.1–1.5
[5]	Piezoresistive	Band-pass filter	0.05–2
[45]	Accelerometer	Low-pass filter	1

Table 1.
Synthesis of the use of filters in respiratory signals.

Reference	Sensor	Objective	Sample rate
[10]	Chest or abdominal belt with a resistive sensor	Talking detection	100 Hz
[25]	Acoustic sensor configured as a headset over the throat	Activity detection of deep breath, eating, drinking, speaking, whispering, whistling, laughing, sighing, and coughing.	22,050 Hz
[3]	Acoustic sensor fixed with tape near the nose	Sleep apnea detection	44.1 kHz
[44]	Acoustic sensor fixed with tape on the thoracic region	Measurement of acoustic sounds from the thorax, including the lung sounds	4 kHz
[23]	Acoustic sensor embedded in a wearable mechanical design over the right chest	Wheeze detection	2048 Hz

Table 2.
Examples of sampling rates.

researchers, therefore, set f_s = 22,050 Hz, which covers up to 11,025 Hz, because the range was considered enough for their application. **Table 2** shows some of the sampling rates used.

Other equally important factors affect the quality of the acquisition, operation, and energy efficiency of wearable devices, such as the duration of acquisition, signal conditioning, conversion resolution, etc. However, these are not explored in this chapter.

4.4 Fast Fourier transform (FFT)

Fast Fourier Transform (FFT) is an algorithm that converts the signal from the time domain to the frequency domain and vice versa [40, 48]. This algorithm is important because it is the first step to extract spectral features, which can be used by machine learning algorithms and other algorithms for signal processing.

5. Machine learning for respiratory signal pattern detection

Machine Learning is the result of pattern recognition and the assumption that computers can learn to execute a task. As a field of artificial intelligence, machine learning is the ability of a machine to learn, identify, and classify from being exposed to specific data in an interactive way, and to not only learn and make reliable decisions but also to adapt when exposed to new data.

This technique can be useful for automatic pattern recognition in respiratory signals such as sleep apnea, respiratory patterns, and talking detection [10, 49, 50]. The steps to implement a machine learning algorithm are introduced in the following sections.

5.1 Feature extraction

First, for machine learning classification, some features must be given to the classification algorithm. These features must be extracted from the original signal, and they must be well chosen for better results.

For example, when working with a wearable acoustic sensor [50] aiming to recognize activity patterns like sitting, eating, and drinking and respiratory patterns such as whispering, deep breath, and coughing, the features extracted from the sensor signals were related to time, frequency, and cepstral:

- Time domain features: these features were obtained using the zero-crossing rate, that is, the rate of sign changes along a signal.

- Frequency domain features: to obtain these features, the FFT needs to be calculated. The features include total spectrum power, subband powers (summed power signal in logarithmically divided bands), brightness (frequency centroid), spectral roll-off (skewness of the spectral distribution), and spectral flux (L2-norm of the spectral amplitude difference of two adjacent frames, representing how drastically the sound changes between two frames).

- Cepstral features: commonly used for speech recognition and audio, the mel-frequency cepstral coefficients are extracted with the application of a discrete cosine transform to the log-scaled outputs of the FFT coefficients filtered by a triangular band-pass filter bank.

It is also possible to use a tool that automatically extracts the features of the signals being studied. With the purpose of identifying talking in respiratory signals [10], more than 10 features were extracted using the Python library "tsfresh" [51]; those that presented more than 10% of recurrence between the tests were manually selected in order to use that feature for classification in the algorithm.

It is common to extract a variety of features in a study, but the effectiveness of a machine learning algorithm strongly depends on which one will be selected and how the data will be selected for training and validation.

5.2 Classification selection

After the selection of features to be used in the algorithm, it is important to decide which the classes are and how the data will be processed. It is important to select what will be used to train the algorithm and what will be used to validate it. There are several ways of separating the acquired data so that the network is trained without the risk of overfitting.

For instance, in Yatani and Tuong [25], two approaches were carried out:

- "Leave-one-participant-out": they worked with 9 samples of data, training the chosen algorithm and using one participant to validate the results.

- "Leave-one-sample-per-participant": an example of each class of each participant was left out for validation and the rest used for training.

A different approach was used by Ejupi and Menon [10]: the data were obtained executing different activities such as walking, standing, and sitting, and an algorithm was trained for each one. For classification, 70% of the database was used for training and 30% for validation.

These techniques prevent the major problem in machine learning, overfitting [52]. In case an algorithm is overfitted, it will produce inaccurate results creating unrealistic patterns. It is always wise to select which data will be used to train the algorithm and which will be used to validate the results, never using all dataset to just one task.

5.3 Machine learning algorithms

The strategy or algorithm to be used in a project as well as its effectiveness and performance are strongly dependent on the problem domain (e.g., data structure, database size, etc.) [53]. It is therefore impossible to choose a method as the best one regardless of domain intricacies. Some popular machine learning algorithms are presented in the following topics.

5.3.1 Support Vector Machines

In order to identify speech pattern using a wearable textile-based sensor [10], the best results were obtained with Support Vector Machines (SVMs). The basic approach for SVM algorithms is to give a set of basic examples and their weight, generally understood as positive and negative (binary) examples for the algorithm, interpreted as classes, where there is a degree of similarity between them, a kernel function, as a means of comparison [52].

SVM was applied for identifying activities using an acoustic sensor [50]. They used more than two classes, comparing one against the other as a strategy to obtain results, using the Radial Basis Function (RBF) as a kernel function. All the implementation using a library "LIBSVM" [54] reaches almost 80% of accuracy.

5.3.2 Naïve Bayes

The Naïve Bayes algorithm can be used when it is necessary to recognize the user activities in real time [25]. The theorem is based on the Bayes statistical theorem that describes the probability of an event based on conditions or previous knowledge. The "naïve" comes from the naivety of the assumption that the results are independent given the cause [52].

From the Bayes' theorem, we have Eq. (6):

$$P(A|B) = \frac{P(B|A) * P(A)}{P(B)} \tag{6}$$

where, P(A|B) is the probability that hypothesis A is true given data of type B. P(B|A) is the probability of data B given that hypothesis A was true.

P(A) is the probability that A is true independently of data, and P(B) is the probability of data B regardless of the hypothesis.

The algorithm uses this probability structure to classify at least two independent sets, which can lead to another set of classification or decision and, at the same time, to another independent set.

This algorithm is simple, computationally cost-effective and can be used for small datasets, as it was used to identify activity patterns such as speaking, laughing, and coughing, presenting good results of accuracy [25].

5.3.3 Artificial neural networks

An artificial neural network (ANN) is a technique based on a series of connected inputs and outputs. Its structure resembles neurons, each one connected and with associated weights. The weights represent information being used by the net to solve the problem and can be adjusted as required. The networks can be supervised or not, the fundamental difference is that in supervised learning, the target vectors indicate what is wanted from the network.

For example, the application of an ANN in talking [10, 25] recognition through respiratory patterns [10] is of supervised learning as the targets to classify are provided to the algorithm.

The neural networks can also be more complex, which depends of the problems intricacies. Aiming to recognize activity patterns such as respiratory effort, using a wearable piezo sensor [25] it was applied networks with up to 17 layers and inputs, a very complex ANN, to achieve the best classification.

Overall, the use of machine learning has become increasingly common in health implementations and has proved a very beneficial tool in classifying and recognizing respiratory activities and patterns when combined with wearable sensors [10, 25, 55].

6. Conclusion

Wearable devices for breathing monitoring and pattern detection are not simple devices. They must not interfere with the respiratory system activities and need to be highly immune to external perturbations. The understanding of the respiratory mechanics is crucial to the development of wearable sensors, and to know how to connect them in an optimal way. The methods, whether manual processing or the use of machine learning algorithms, depend substantially on the type of the signal studied and are a crucial step for a study development.

Acknowledgements

The authors thank CAPES (Coordenação de Aperfeiçoamento de Pessoal de Nível Superior), Fundação Araucária de Apoio ao Desenvolvimento Científico e Tecnológico do Estado do Paraná, and CNPq (Conselho Nacional de Desenvolvimento Científico e Tecnológico), Brazil, for the scholarships and the financial support.

Conflict of interest

The authors declare no conflict of interest.

Author details

Taisa Daiana da Costa[1], Maria de Fatima Fernandes Vara[2], Camila Santos Cristino[2], Tyene Zoraski Zanella[2], Guilherme Nunes Nogueira Neto[2] and Percy Nohama[1,2*]

1 Universidade Tecnológica Federal do Paraná, Curitiba, Paraná, Brazil

2 Pontifícia Universidade Católica do Paraná, Curitiba, Paraná, Brazil

*Address all correspondence to: percy.nohama@gmail.com

References

[1] Aliverti A. Wearable Technology: Role in Respiratory Health and Disease. Breathe [Internet]. 2017 Jun 18;**13**(2):e27-e36. Available from: http://breathe.ersjournals.com/lookup/doi/10.1183/20734735.008417

[2] Dias D, Cunha JPS. Wearable health devices—vital sign monitoring, systems and technologies. Sensors (Switzerland). 2018;**18**(8):1-28

[3] Fang Y, Jiang Z, Wang H. A Novel Sleep Respiratory Rate Detection Method for Obstructive Sleep Apnea Based on Characteristic Moment Waveform. Journal of Healthcare Engineering [Internet]. 2018;**2018**:1-10. Available from: https://www.hindawi.com/journals/jhe/2018/1902176/

[4] Pang Y, Jian J, Tu T, Yang Z, Ling J, Li Y, et al. Wearable Humidity Sensor Based on Porous Graphene Network for Respiration Monitoring. Biosensors and Bioelectronics [Internet]. 2018 Sep;116(March):123-129. Available from: https://doi.org/10.1016/j.bios.2018.05.038

[5] Molinaro N, Massaroni C, Lo Presti D, Saccomandi P, Di Tomaso G, Zollo L, et al. Wearable Textile Based on Silver Plated Knitted Sensor for Respiratory Rate Monitoring. In: 2018 40th Annual International Conference of the IEEE Engineering in Medicine and Biology Society (EMBC) [Internet]. IEEE. 2018:2865-2868. Available from: https://ieeexplore.ieee.org/document/8512958/

[6] Steinhubl SR, Muse ED, Topol EJ. The Emerging Field of Mobile Health. Science Translational Medicine [Internet]. 2015 Apr 15;7(283):283rv3-283rv3. Available from: http://stm.sciencemag.org/lookup/doi/10.1126/scitranslmed.aaa3487

[7] Dieffenderfer J, Goodell H, Mills S, McKnight M, Yao S, Lin F, et al. Low-power Wearable Systems for Continuous Monitoring of Environment and Health for Chronic Respiratory Disease. IEEE Journal of Biomedical and Health Informatics [Internet]. 2016 Sep;**20**(5):1251-1264. Available from: http://ieeexplore.ieee.org/document/7479442/

[8] Ionescu CM, Copot D. Monitoring Respiratory Impedance by Wearable Sensor Device: Protocol and Methodology. Biomedical Signal Processing and Control [Internet]. 2017 Jul;**36**:57-62. Available from: http://dx.doi.org/10.1016/j.bspc.2017.03.018

[9] Ho K, Paez J, Liu B. Air Quality Alerts Benefit Asthmatics. The Lancet Planetary Health [Internet]. 2018 Dec 1 [cited 2019 Feb 4];**2**(12):518. Available from: http://www.ncbi.nlm.nih.gov/pubmed/30526931

[10] Ejupi A, Menon C. Detection of Talking in Respiratory Signals: A Feasibility Study Using Machine Learning and Wearable Textile-Based Sensors. Sensors [Internet]. 2018 Jul 31;**18**(8):2474. Available from: http://www.mdpi.com/1424-8220/18/8/2474

[11] Ernst F, Saß P. Respiratory Motion Tracking Using Microsoft's Kinect v2 Camera. Current Directions in Biomedical Engineering [Internet]. 2015;**1**(1):192-195. Available from: http://www.degruyter.com/view/j/cdbme.2015.1.issue-1/cdbme-2015-0048/cdbme-2015-0048.xml

[12] De Troyer A. Effect of hyperinflation on the diaphragm. The European Respiratory Journal. 1997;**10**(3):708-713

[13] Tortora GJ, Derrickson B. Principles of Anatomy & Physiology. 14th ed. Wiley; 2014. p. 1232

[14] Grigor Abrahamyan M. Some Aspects of the Physics of Gas Flow in

the Respiratory System. International Journal of Clinical and Experimental Medical Sciences [Internet]. 2018;**4**(1):1. Available from: http://www.sciencepublishinggroup.com/journal/paperinfo?journalid=335&doi=10.11648/j.ijcems.20180401.11

[15] Lehrer S. Understanding Lung Sounds: Third Edition [Internet]. CreateSpace Independent Publishing Platform; 2002. p. 158. Available from: https://books.google.com.br/books?id=eFFUDwAAQBAJ

[16] Caruso P, Albuquerque ALP de, Santana PV, Cardenas LZ, Ferreira JG, Prina E, et al. Diagnostic Methods to Assess Inspiratory and Expiratory Muscle Strength. Jornal Brasileiro de Pneumologia [Internet]. 2015 Apr;**41**(2):110-123. Available from: http://www.scielo.br/scielo.php?script=sci_arttext&pid=S1806-37132015000200110&lng=en&tlng=en

[17] Liu H, Guo S, Liu H, Zhang H. The best body spot to detect the vital capacity from the respiratory movement data obtained by the wearable strain sensor. Journal of Physical Therapy Science. 2018;586-589

[18] Johnston CR, Krishnaswamy N, Krishnaswamy G. The Hoover's Sign of Pulmonary Disease: Molecular Basis and Clinical Relevance. Clinical and Molecular Allergy [Internet]. 2008 Dec 5;**6**(1):8. Available from: https://clinicalmolecularallergy.biomedcentral.com/articles/10.1186/1476-7961-6-8

[19] Maestri R, Robbi E, Lovagnini M, Bruschi C, La Rovere MT, Pinna GD. Arterial Oxygen Saturation During Cheyne-Stokes Respiration in Heart Failure Patients: Does Measurement Site Matter? Sleep Medicine [Internet]. 2018 Dec; Available from: https://linkinghub.elsevier.com/retrieve/pii/S1389945718302399

[20] Fouzas S, Anthracopoulos MB, Bohadana A. Clinical Usefulness of Breath Sounds. In: Breath Sounds [Internet]. Cham: Springer International Publishing; 2018. p. 33-52. Available from: http://link.springer.com/10.1007/978-3-319-71824-8_3

[21] Bohadana A, Izbicki G, Kraman SS. Fundamentals of Lung Auscultation. The New England Journal of Medicine. 2014;**370**(8):744-751

[22] Sarkar M, Madabhavi I, Niranjan N, Dogra M. Auscultation of the Respiratory System. Annals of Thoracic Medicine [Internet] 2015;**10**(3):158. Available from: http://www.thoracicmedicine.org/text.asp?2015/10/3/158/160831

[23] Li S-H, Lin B-S, Tsai C-H, Yang C-T, Lin B-S. Design of Wearable Breathing Sound Monitoring System for Real-Time Wheeze Detection. Sensors [Internet]. 2017 Jan 17;**17**(12):171. Available from: http://www.mdpi.com/1424-8220/17/1/171

[24] Reinvuo T, Hannula M, Sorvoja H, Alasaarela E, Myllyla R. Measurement of Respiratory Rate with High-Resolution Accelerometer and Emfit Pressure Sensor. In: Proceedings 2006 IEEE Sensors Applications Symposium, 2006 [Internet]. IEEE; 2006 [cited 2015 Jan 6]. 192-195. Available from: http://ieeexplore.ieee.org/lpdocs/epic03/wrapper.htm?arnumber=1634270

[25] Yatani K, Truong KN. BodyScope. In: Proceedings of the 2012 ACM Conference on Ubiquitous Computing - UbiComp '12 [Internet]. New York, New York, USA: ACM Press; 2012. 341. Available from: http://dl.acm.org/citation.cfm?doid=2370216.2370269

[26] Yuasa Y, Takahashi K, Suzuki K. Wearable Flexible Device for Respiratory Phase Measurement Based on Sound and Chest Movement. 2017 IEEE International

Conference on Systems, Man, and Cybernetics (SMC). 2017;2017–Jan:2378-2383

[27] Shkel AA, Kim ES. Wearable Low-power Wireless Lung Sound Detection Enhanced by Resonant Transducer Array for Pre-filtered Signal Acquisition. In: 2017 19th International Conference on Solid-State Sensors, Actuators and Microsystems (TRANSDUCERS) [Internet]. IEEE; 2017. 842-845. Available from: http://ieeexplore.ieee.org/document/7994180/

[28] Jubran A. Pulse Oximetry. Critical Care [Internet]. 2015 Dec 16 [cited 2019 Feb 2];**19**(1):272. Available from: http://ccforum.com/content/3/2/R11

[29] Jortberg E, Silva I, Bhatkar V, McGinnis R, Sen-Gupta E, Morey B, et al. A Novel Adhesive Biosensor System for Detecting Respiration, Cardiac, and Limb Movement Signals During Sleep: Validation with Polysomnography. Nature and Science of Sleep [Internet]. 2018 Nov;**10**:397-408. Available from: https://www.dovepress.com/a-novel-adhesive-biosensor-system-for-detecting-respiration-cardiac-an-peer-reviewed-article-NSS

[30] Moussavi ZK, Leopando MT, Pasterkamp H, Rempel G. Computerised acoustical respiratory phase detection without airflow measurement. Medical & Biological Engineering & Computing. 2000;**38**:198-203

[31] Amini N, Sarrafzadeh M, Vahdatpour A, Xu W. Accelerometer-based On-body Sensor Localization for Health and Medical Monitoring Applications. Pervasive and Mobile Computing [Internet]. 2011 Dec [cited 2019 Feb 2];**7**(6):746-760. Available from: https://www.ncbi.nlm.nih.gov/pmc/articles/PMC3279922/pdf/nihms328466.pdf

[32] Jin A, Yin B, Morren G, Duric H, Aarts RM. Performance Evaluation of a Tri-axial Accelerometry-based Respiration Monitoring for Ambient Assisted Living. In: 2009 Annual International Conference of the IEEE Engineering in Medicine and Biology Society [Internet]. IEEE; 2009. 5677-5680. Available from: http://ieeexplore.ieee.org/document/5333116/

[33] Wu Y. Novel High Sensitivity Accelerometer based on a Microfiber Loop Resonator. Optical Engineering [Internet]. 2010 Jan 1;**49**(1):014402. Available from: http://opticalengineering.spiedigitallibrary.org/article.aspx? doi: 10.1117/1.3294883

[34] Gomathi T, Shaby SM. Capacitive Accelerometers for Microelectromechanical Applications: A Review. In: 2016 International Conference on Control, Instrumentation, Communication and Computational Technologies (ICCICCT) [Internet]. IEEE; 2016. 486-490. Available from: http://ieeexplore.ieee.org/document/7987999/

[35] Hung PD, Bonnet S, Guillemaud R, Castelli E, Yen PTN. Estimation of Respiratory Waveform Using an Accelerometer. In: 2008 5th IEEE International Symposium on Biomedical Imaging: From Nano to Macro [Internet]. IEEE; 2008. 1493-1496. Available from: http://ieeexplore.ieee.org/document/4541291/

[36] Rendon DB, Ojeda JLR, Foix LFC, Morillo DS, Fernandez MA. Mapping the Human Body for Vibrations using an Accelerometer. In: 2007 29th Annual International Conference of the IEEE Engineering in Medicine and Biology Society [Internet]. IEEE; 2007. 1671-1674. Available from: http://ieeexplore.ieee.org/document/4352629/

[37] Burnett TA, Mann EA, Stoklosa JB, Ludlow CL. Self-Triggered Functional Electrical Stimulation During Swallowing. Journal of Neurophysiology [Internet]. 2005 Dec;**94**(6):4011-4018.

Available from: http://www.physiology.org/doi/10.1152/jn.00025.2005

[38] Zięba J, Frydrysiak M. Textronics – Electrical and electronic textiles. Sensors for breathing frequency measurement. Fibres & Textiles in Eastern Europe. 2006;**14**(5):43-48

[39] Mitchell E, Coyle S, O'Connor NE, Diamond D, Ward T. Breathing Feedback System with Wearable Textile Sensors. In: 2010 International Conference on Body Sensor Networks [Internet]. IEEE; 2010. 56-61. Available from: http://ieeexplore.ieee.org/document/5504719/

[40] Antoniou A. Digital Signal Processing: Signals, Systems and Filters [Internet]. McGraw-Hill; 2006. p. 965. Available from: https://books.google.com.br/books?id=JQ4fAQAAIAAJ

[41] Zhao T, Liu X, Zhang G, Su Y. Design of a Programmable and Low-frequency Filter for Biomedical Signal Sensing Applications. In: 2016 9th International Congress on Image and Signal Processing, BioMedical Engineering and Informatics (CISP-BMEI) [Internet]. IEEE; 2016. p. 1746-1750. Available from: http://ieeexplore.ieee.org/document/7852999/

[42] RM R. Introduction to Biomedical Signals. In: Biomedical Signal Analysis. John Wiley & Sons; 2015

[43] Corbishley P, Rodriguez-Villegas E. Breathing Detection: Towards a Miniaturized, Wearable, Battery-Operated Monitoring System. IEEE Transactions on Biomedical Engineering [Internet]. 2008 Jan [cited 2015 Jan 6];**55**(1):196-204. Available from: http://www.ncbi.nlm.nih.gov/pubmed/18232362

[44] Kirchner J, Souilem S, Fischer G. Wearable System for Measurement of Thoracic Sounds with a Microphone Array. In: 2017 IEEE Sensors Journal [Internet]. IEEE; 2017. 1-3. Available from: http://ieeexplore.ieee.org/document/8234248/

[45] Bates A, Ling MJ, Mann J, Arvind DK. Respiratory Rate and Flow Waveform Estimation from Tri-axial Accelerometer Data. In: 2010 International Conference on Body Sensor Networks [Internet]. IEEE; 2010 [cited 2014 Dec 25]. 144-150. Available from: http://ieeexplore.ieee.org/lpdocs/epic03/wrapper.htm?arnumber=5504743

[46] Shannon CE. A Mathematical Theory of Communication. Bell System Technical Journal [Internet]. 1948 Jul;**27**(3):379-423. Available from: http://ieeexplore.ieee.org/lpdocs/epic03/wrapper.htm?arnumber=6773024

[47] Nyquist H. Nyquist-Shannon Sampling Theorem. Princeton University; 2012

[48] Walker JS. Fast Fourier Transforms. CRC Press; 2017. p. 464. (Studies in Advanced Mathematics)

[49] McCaughey EJ, McLachlan AJ, Gollee H. Non-Intrusive Real-time Breathing Pattern Detection and Classification for Automatic Abdominal Functional Electrical Stimulation. Medical Engineering & Physics [Internet]. 2014;**36**(8):1057-1061. Available from: http://dx.doi.org/10.1016/j.medengphy.2014.04.005

[50] Yatani K, Truong KN. BodyScope: A Wearable Acoustic Sensor for Activity Recognition. In: Ubicomp '12 The 2012 ACM Conference on Ubiquitous Computing. Pittsburgh, PA, USA; 2012. 341-350

[51] Pedregosa F, Weiss R, Brucher M. Scikitlearn: Machine learning in Python. Journal of Machine Learning Research. 2011;**12**:2825-2830

[52] Domingos P. The Master Algorithm: How the Quest for the Ultimate Learning Machine Will Remake Our World. 2015. p. 316

[53] Hernández-Pereira EM, Álvarez-Estévez D, Moret-Bonillo V. Automatic Classification of Respiratory Patterns Involving Missing Data Imputation Techniques. Biosystems Engineering [Internet]. 2015 Oct;138:65-76. Available from: https://linkinghub.elsevier.com/retrieve/pii/S1537511015001117

[54] Chang C, Lin C, Tieleman T. LIBSVM : A Library for Support Vector Machines. ACM Transactions on Intelligent Systems and Technology. 2008;**307**:1-39

[55] Parkka J, Ermes M, Korpipaa P, Mantyjarvi J, Peltola J, Korhonen I. Activity Classification Using Realistic Data From Wearable Sensors. IEEE Transactions on Information Technology in Biomedicine [Internet]. Jan 2006;**10**(1):119-128. Available from: http://ieeexplore.ieee.org/document/1573714/

Section 2

Education and Social Interactions

Chapter 5

Wearable Electromechanical Sensors and Its Applications

Dan Liu and Guo Hong

Abstract

Wearable electromechanical sensor transforms mechanical stimulus into electrical signals. The main electromechanical sensors we focus on are strain and pressure sensors, which correspond to two main mechanical stimuli. According to their mechanisms, resistive and capacitive sensor attracts more attentions due to their simple structures, mechanisms, preparation method, and low cost. Various kinds of nanomaterials have been developed to fabricate them, including carbon nanomaterials, metallic, and conductive polymers. They have great potentials on health monitoring, human motion monitoring, speech recognition, and related human-machine interface applications. Here, we discuss their sensing mechanisms and fabrication methods and introduce recent progress on their performances and applications.

Keywords: wearable, electromechanical sensor, health monitoring, fabrication, mechanism

1. Introduction

With the rapid development of information technology, the Internet of Everything turns more critical in the next technological revolution. Wearable devices, which have the advantages of good portability, easy to carry, and multi-functional capability, are considered as the basic hardware in the future, which show great potential on many applications, including medicine, healthcare, robotic systems, prosthetics, visual realities, professional sports, as well as entertainment. In recent years, much efforts have been devoted to developing wearable sensing technologies. Various kinds of wearable sensors have been proposed and demonstrated in lab, from single functional sensors, such as temperature [1], pressure [2], strain [3], optical [4], and electrochemical sensors [5], to multifunctional sensors, such as tactile and electronic skin [6]. Among these wearable sensors, wearable electromechanical sensors including strain and pressure sensor have attracted more and more attentions due to its clear mechanism, low cost, low power consumption, and high performance [7]. Through integrating wearable strain and pressure sensor with other sensors, tactile sensor [8] and electronic skin [9] have been realized. High-performance wearable electromechanical sensor can monitor the tiny change of strain and pressure, which is useful in many fields.

Traditional electromechanical sensor is usually fabricated with brittle materials, such as silicon and metal. Though flexibility can be improved by structural design, their performance is still limited. Thus, many new materials have been developed. The materials used in wearable electromechanical sensor consist of sensing and supporting material. Most of the progresses are focusing on the development of

IntechOpen

new sensing materials. Structural design is also an effective strategy to improve the performance. Fabrication method is also the significant aspect. Many traditional techniques are utilized, such as screen printing, contact printing, electrospinning, and spray coating [10]. Moreover, wearable electromechanical sensor has been successfully demonstrated on a lot of applications, such as health monitoring, disease diagnosis, behavior correction, alarm of accident falls, human-machine interfaces, and even speech recognition.

The present chapter will discuss their basic working mechanism, fabrication methods, and applications of wearable electromechanical sensors and challenges facing the progress.

2. Working mechanisms of a wearable electromechanical sensor

Firstly, we discuss the working mechanism of a wearable electromechanical sensor. Based on their working mechanisms, it can be classified into piezoresistive, capacitive, iontronic, and piezoelectric sensor, as seen in **Figure 1** [11].

2.1 Wearable piezoresistive sensor

Figure 1a shows the mechanism of piezoresistive sensor. It transfers mechanical stimuli into resistance signal. The factors resulting in resistance change depend on the property of materials utilized and their structures, including geometrical effect, structural effect, and disconnection mechanism.

2.1.1 Geometrical effect

Geometrical effect means that the resistance change is caused by geometrical change, which is mainly due to Poisson's ratio (υ). Poisson's ratio (υ) is a fundamental parameter of materials, meaning that materials tend to contract in transverse direction of stretching when they are stretched. The resistance of a conductor is represented by:

$$R = \rho L / A \quad (1)$$

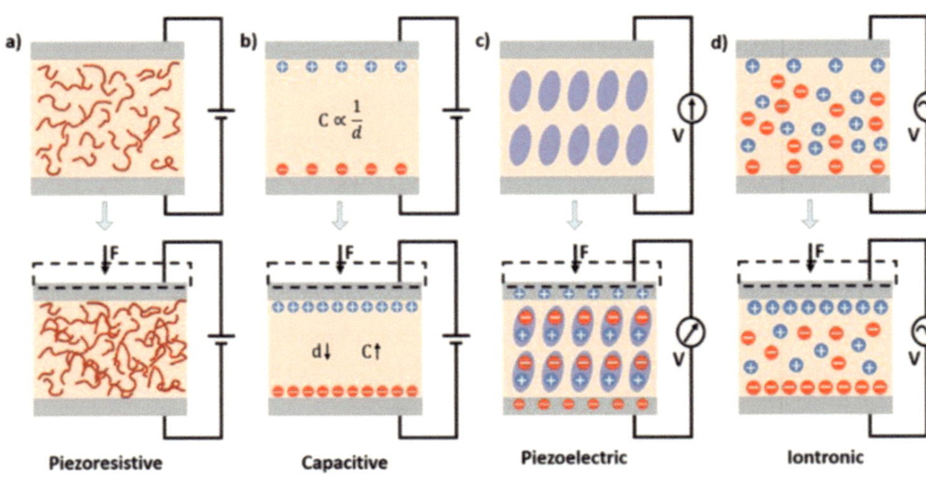

Figure 1.
Schematics illustrating the different modalities of wearable electromechanical sensors. (a) Piezoresistivity, (b) capacitance, (c) piezoelectricity, and (d) iontronic.

where ρ is the electrical resistivity, L is the length, and A is the cross-sectional area of the conductor. When strain or pressure is applied, the length increases and cross-sectional area would be changed due to the shrinkage of materials, resulting in change of the resistance. Geometrical effect is usually limited compared to other factors.

2.1.2 Structural effect

Structural effect is defined as the change in the resistance caused by the structural deformations. This is usually observed in semiconducting materials. When strain or pressure is applied, the crystal structure especially interatomic space is changed, resulting in the change of the bandgap, which may increase the resistance of materials to few times [12]. For example, individual carbon nanotube (CNT) [13] shows ultrahigh resistivity change owning to their chirality and change in barrier height, respectively. However, compared with total resistance change, the part is usually low because strain applied on individual nanoflake is always small. In addition, the large elastic mismatch and weak interfacial adhesion strength between nanomaterials and polymers also make nanoflakes almost free from deformation.

2.1.3 Disconnection mechanism

The disconnection mechanism means that resistance change is caused by disconnection process between adjacent nanoflakes. It consists of three situations under different strains or pressures, which are contact area change, tunneling effect, and crack propagation.

When the applied strain or pressure is small, contact area changes between adjacent nanoflakes dominants. The electrons mainly pass through overlapped nanoflakes within the percolation conductive network. When the applied strain or pressure increases and fully pull some adjacent nanoflakes apart, the electrons still can pass through them because the distance between them is small enough. This phenomenon is called tunneling effect, and the distance is called tunneling distance. The tunneling resistance between two adjacent nanoflakes can be approximately estimated by Simmons's theory [14]:

$$R_{tunnel} = \frac{h^2 d}{Ae^2\sqrt{2m\lambda}} \exp\left(\frac{4\pi d}{h}\sqrt{2m\lambda}\right) \quad (2)$$

where A, e, h, d, m, λ represent the cross-sectional area of the tunneling junction, single electron charge, Plank's constant, the distance between adjacent nanoflakes, the mass of electron and the height of energy barrier for insulators, respectively. It can be found that the distance between adjacent nanoflakes dominates the tunneling resistance. When there is no electron pass through by tunneling, the distance is defined as cut-off tunneling distance. The cut-off distance is usually several nanometers. When the applied strain or pressure is large enough, crack is formed, leading to rapidly increasing of resistance. Strain or pressure leads opening and enlargement of cracks, critically limiting the electrical conduction due to the separation of several crack edges.

2.2 Wearable capacitive sensor

As **Figure 1b** shows wearable capacitive sensor is based on capacitance change of capacitor. Among different capacitors, the most popular architecture is the parallel-plate configuration because it is easy to be fabricated and its model is simple. The capacitive change can be expressed by the classic equation:

$$C = \kappa \frac{A}{d} \tag{3}$$

in which κ, A, and d represent the permittivity of the medium between two plates, the overlap area, and the distance between two plates, respectively. When any of them is changed by the mechanical stimulus, the capacitance would be changed.

For capacitive strain sensor, when the strain ε is applied, the length of capacitor along the strain direction would be increased, which is expressed as $(1 + \varepsilon)l_0$, while the width and thickness of dielectric layer would be decreased, which is expressed as $(1 - \nu_{electrode})w_0$ and $(1 - \nu_{dielectric})d_0$, respectively. The $\nu_{electrode}$ and $\nu_{dielectric}$ are used to represent the Poisson's ratios of flexible electrodes and dielectric layer, respectively. If both flexible electrodes and dielectric layer have same Poisson's ratio, then the capacitance upon stretching could be calculated as:

$$C = (1 + \varepsilon)\, C_0 \tag{4}$$

The equation indicates that the capacitance of capacitive strain sensor is linear with the applied strain. However, the linear relationship is only suitable for limited strain range. When the applied strain is higher than certain value, the relationship between different axes cannot be obtained simply by the Poisson's ratio.

For capacitive pressure sensor, the sensitivity (S) of capacitance to pressure is given by:

$$S = \delta(\Delta C / C_0) / \delta P \tag{5}$$

where ΔC is the variation of capacitance ($C\text{–}C_0$) and P presents applied pressure. The most popular structure for the wearable pressure sensor is interlock structure, which is hard to make accurate analysis.

2.3 Iontronic sensors

As **Figure 1c** shows, iontronic sensor is based on the iontronic interface sensing mechanism. The iontronic interface usually exists at the nanoscale interface between the electrode and the electrolyte. The electrode forms ionic-electronic contact with ionic gel. The electrons on the electrode and the counter ions from the iontronic film accumulate and attract to each other at a nanoscopic distance, leading to an ultrahigh unit-area capacitance. Compared to traditional parallel plate capacitive sensors, iontronic sensor has a higher surface area and its electrical capacitance is at last 1000 times larger. This excellent property is suitable for wearable electromechanical sensors. In addition, this special mechanism enables iontronic sensor immunity to environmental or body capacitive noises. So far, ion gels and ionic liquids are the most popular materials for iontronic sensor.

2.4 Piezoelectric sensors

As **Figure 1d** shows, the sensing mechanism of piezoelectric sensor is piezoelectric effect. Piezoelectric means that electric change accumulates in piezoelectric materials when mechanical stress is applied. Many materials have piezoelectric property, such as crystals, certain ceramics, and even biological matter. When strain or pressure is applied, there is a change in electrical polarization inside the material, resulting in a change in surface charge (voltage) at the surface of the piezoelectric material. In general, the electrical signal of piezoelectric sensor is voltage, which can be collected by measuring two different surfaces.

3. Performance of wearable electromechanical sensor

3.1 Basic parameters of wearable electromechanical sensor

3.1.1 Sensitivity and linearity

Sensitivity is the magnitude of electrical response to measured mechanical stimulus, which is an important parameter. For strain sensor, sensitivity is called gauge factor (GF), which is defined as GF = $\Delta R/R_0$ for resistive type and GF = $\Delta C/C_0$ for capacitive type. For pressure sensor, pressure sensitivity (PS) is defined as PS = $(\Delta R/R_0)/P$. Sensitivity can be affected by functional material, sensing mechanism, and structural configuration. The materials with large piezoresistive or piezoelectric coefficient are desired. Tunneling effect and crack/gap structures in piezoresistive sensors have been proven to be effective in promoting sensitivity. However, most highly sensitive sensors always show limited stretchability.

Linearity characterizes degree of deviation from linear relationship between electrical signals and mechanical stimulus. High linearity is convenient for the calibration and data processing process. However, there is always a contradiction between sensitivity and linearity because crack propagation and tunneling-effect-induced resistance change are usually exponential. For instance, piezoresistive strain sensors often exhibit varied sensitivity in different strain ranges, which is induced by the nonlinear heterogeneous deformation. In addition, capacitive sensors with microstructured dielectric also suffer the similar problem.

3.1.2 Hysteresis and response time

Hysteresis and response time are another two important parameters in evaluating dynamical performance of electromechanical sensor. Hysteresis means the dependence of the performance on its history, which should be reduced or avoided. In general, capacitive sensors show immediate responding to the variation of overlapped area, featuring a lower hysteresis. Meanwhile, piezoresistive sensors have slower response due to the interactive motion between sensing material and polymer substrate. The interfacial binding between sensing material and substrate greatly affects the optimization of hysteresis. The full recovery of sensing material position is hindered by the interfacial slide, leading to a high hysteresis behavior. Meanwhile, to avoid the friction-induced buckling and facture in sensing materials, a weak adhesion is needed. It is reported that using low viscoelastic polymer substrate and improved configuration can partially eliminate hysteresis. However, it is still a large challenge to optimize hysteresis by novel material and structural engineering. Response time illustrates the speed to achieve steady response to applied mechanical stimulus, and response delay exists in nearly all composite-based sensors because of the viscoelastic property of polymers. Relatively, piezoresistive device has a larger response time than others because it needs more time to reestablish percolation network in resistive composites. In addition, lower modulus materials are popular for wearable electromechanical sensor, which can further decrease the response speed of resistive sensors. Moreover, based on structural design, the newly developed crack-based piezoresistive sensors show an appealing response time (about 20 ms) because cracks can reversibly connect and disconnect with loading and unloading of mechanical stimuli [15].

3.1.3 Durability

Durability is the ability to remain its performance, without requiring excessive maintenance or repair, when it is normally used. It is usually measured by cyclic stability for wearable electromechanical sensor. Cyclic stability is sensor endurance to periodic loading and unloading cycles. The sensing material film on polymer substrate is easy to form buckling, facture, and even stripping after enough cycles, which results in cyclic instable problem. For example, the sensitivity of graphene woven fabric (GWF) strain sensor decreases 24% after about 1000 cycles from 0 to 2% [16].

Endowing sensor with self-healing is a novel way to promoting durability. Several works have been reported on wearable electromechanical sensor. **Figure 2a** shows a stretchable self-healing piezoresistive strain sensor using single wall carbon nanotube (SWCNT) in self-healing hydrogel (SWCNT/hydrogel) as the conductive sensing channel [17]. The cutting groove is partially healed after 30 s and totally restored to normal after 60 s at room temperature without any external assistance. It also shows the repetitive cutting-healing processes with five cycles at the same location. The average efficiencies are 98 ± 0.8% for the five self-healing cycles within about 3.2 s, indicating that the SWCNT/hydrogel possesses significant and repeatable electrical restoration performance. **Figure 2b** shows that a self-healing sensor with tunable positive/negative piezoresistivity is designed by the construction of hierarchical structure connected through supramolecular metal-ligand coordination bonds [18]. The electrical resistance of the repaired samples only slightly increases after multiple cutting/healing cycles. However, the increase of electrical resistance is neglectable, which is lower than one order of magnitude, indicating its excellent electrical self-healing ability. The high-healing efficiency is estimated to be 88.6% after the third healing process, and the healed wearable strain sensor still show good flexibility, high sensitivity, and accurate detection capability, even after bending over 10,000 cycles.

3.1.4 Biocompatibility

Wearable electromechanical sensors are usually directly used on human skins, so biocompatibility is also important. The main danger comes from sensing materials, which is usually nanomaterial other than substrate materials, which is a polymer. For example, it has been reported that injecting large quantities of CNTs into mice lungs could cause asbestos-like pathogenicity because of the small size and needle-like morphology of CNT [19]. To improve the biocompatibility, organic active materials, such as polypyrrole (PPy) and poly(3,4-ethylenedioxythiophene)

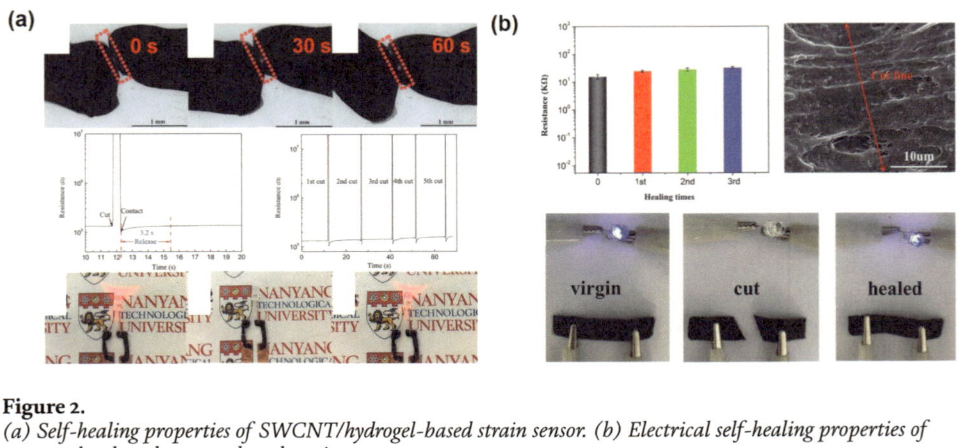

Figure 2.
(a) Self-healing properties of SWCNT/hydrogel-based strain sensor. (b) Electrical self-healing properties of supramolecular-elastomer-based strain sensor.

(PEDOT), have generally been used. The carbonized cotton or silk also presents great potentials in constructing biocompatible wearable sensors [20].

3.1.5 Self-power

Power is the basic element for wearable system. Wearable devices with self-power ability attract more and more attentions, which can greatly extend their application scenarios and is particularly suitable for long-lasting wearables. Self-power wearable electromechanical sensor has been demonstrated so far using triboelectric [21], photovoltaic [22], piezoelectric [23], radiofrequency, thermoelectric (TE) systems [24], and others [25]. Among them, TE technology is rather attractive because of the utilization of conjugated polymers as the active component, which is also flexible, enabling a new generation of novel, low-cost, low-powered wearable electromechanical sensors [26].

3.2 Materials for wearable electromechanical sensor

3.2.1 Materials for substrate

Substrate is mainly responsible for flexibility and stretchability, and directly determines the comfort level and long-term reliability. Polydimethylsiloxane (PDMS), a commercial silicone elastomer with intrinsic high stretchability (up to 1000%), nontoxic, nonflammability, hydrophobicity, and good processability, has been frequently used. Though cannot be stretched for its relatively high modulus (about 2~4 GPa), polyethylene terephthalate (PET) features good transparency (>85%), high creep resistance, and excellent printability. Silicone elastomers including Ecoflex, Sylgard, Dragon Skin, and Silbione are biocompatible and their maximum stretchability is up to 900%. They are suitable flexible substrate because of their strong adhesion onto target surfaces. Ecoflex® rubber is a newly developed, highly stretchable and skin safe silicone with better stretchability and lower modulus, which has been used in the sensors requiring more severe flexibility and stretchability. Polyimide (PI) is another frequently used substrate because it can maintain flexibility, creep resistance and tensile strength under the condition of high temperature (up to 360°C) and acids/alkalis. Thus, PI is compatible with micromanufacturing process and many types of wearable electromechanical sensor are possible to be designed and implemented on it. Natural materials are also explored and developed to produce flexible substrate because they are easily biodegraded, such as cellulose paper. Moreover, the natural textiles, like silk and cotton, are also highly desirable substrate materials [41].

3.2.2 Materials for active elements

3.2.2.1 Carbon nanomaterials

Carbon nanomaterials including graphite, CNT and graphene, have been widely used in fabricating wearable electromechanical sensors. Graphite is a conductor and attracts more and more attentions with development of pencil-on-paper electronics [54]. Graphite flakes in pencil lead is easy to be deposited on paper surface by the physical friction between lead tip and porous cellulose paper. Moreover, structural edges in graphite flakes results in a strain-induced resistance variation of pencil traces, making them suitable for strain sensor. The contact area between graphite flakes increases by compressing the trace and decreases when the tension strain is applied, leading to the decrease or increase of resistance. The wearable strain sensor fabricated with pencil-on-paper shows high GF up to 536.61 [27].

CNT are allotropes of carbon with a cylindrical nanostructure, which possesses excellent electrical conductivity and mechanical properties. It has been demonstrated that a single CNT shows strong structural effect and has a GF higher than 1000. However, wearable electromechanical sensor fabricated with single CNT is difficult and hard to realize mass production. Thus, CNT is usually intermingled into polymer substrates and its excellent conductivity plays an important role in electromechanical sensor construction. Wearable capacitive and piezoresistive electromechanical sensors have all been demonstrated by depositing CNT onto substrate or forming composite with polymers. For the piezoresistive composite sensor, the resistance change is mainly due to the strain-varied intertube tunneling resistance. The maximum GF can be achieved when the concentration of CNT is near the percolation threshold (PH). When the CNT loading is much lower than PH, the distance between adjacent CNTs is larger than their cut-off distance and there is almost no tunneling resistance. On the contrary, when CNT loading is much higher than PH, the CNTs can form dense 3D network and most of CNTs would connect with each other, resulting in a small intertube resistance. In this case, the contact resistance dominates the behavior, which will significantly decrease the GF. For piezoresistive film sensor, the variation of resistance gains almost a tenfold increment compared with nanocomposite type, but its cycle durability is not favorable enough because of unexpected cracks and desquamations. CNTs are also used to form wrinkle structure on a soft substrate via heating of the film or a prestrained substrate and are utilized to fabricate high-performance wearable strain sensor.

Due to outstanding electroconductibility, excellent mechanical properties, great thermal characteristic and optical transmittance, graphene becomes the most promising sensing material for the development of wearable electromechanical sensor [28]. Graphene has been developed as electrode material for capacitive sensor and filler for piezoresistive sensor. A variety of graphene electromechanical sensors with different forms have been demonstrated, including porous foams, flakes, ripples, woven fabrics, and films. For example, the GWF film, which can be fabricated either by CVD or dip coating, consisted of many overlapping microribbons and features a good trade-off between sensitivity and stretchability, making it suitable for wearable strain sensors. It shows fascinating stretchability (a tolerable strain up to 57%) and sensitivity (GF = 416 for $0 < \varepsilon < 40\%$, and GF = 3667 for $48 < \varepsilon < 57\%$) by encapsulating the obtained GWFs in natural rubber latex [29].

3.2.2.2 Metal materials

Metal possesses excellent electrical conductivity and has been widely used in wearable electromechanical sensors. There are four forms of metal developed, which are nanowires, nanoparticles, stretchable configurations, and liquid state at room temperature. Nanowires (NWs) and nanoparticles (NPs) are usually used to prepare piezoresistive composites or conductive ink. For example, silver nanowire (AgNW) can be embedded into PDMS to build resistive-type strain sensor. Because the adhesion between AgNWs and polymers is not as strong as carbon nanomaterials, AgNW interconnection is easy to be broken. The resistance will irreversibly increase after buckling and wrinkling if the AgNW film is just simply coated on the surface of polymer. In addition, AgNWs are easy to be oxidized. Therefore, AgNW layer is often sandwiched between two polymer layers, ensuring AgNWs to move back along their determined paths and be free from oxidation [23]. The stretchable configurations of metal are on the basis of the strategy "structures that are flexible and stretch." Coiled buckled, serpentine and woven structures have been utilized to endow flexibility and stretchability to metals. The liquid metal, like Ga and its alloys, maintains the liquid state at room temperature. With the help of microfluidic techniques, liquid metals show a great potential on wearable sensors. When strain or pressure is applied, the microchannel geometry will be changed, leading to a significant variation in the sectional area and length of liquid metal resistor. The change of electric resistance can reach as much as 50%.

3.2.2.3 Polymer

Conductive polymers possess favorable electroproperties and can participate in building sensing materials. An attractive feature of conductive polymer is the mechanical similarity between them and many insulated substrate polymers. PEDOT-based polymers are the most common sensing materials for their thermal stability, high transparency, and tunable conductivity. Among them, poly(3,4-ethylenedioxythiophene)-polystyrene sulfonate (PEDOT:PSS) is one of the promising conductive polymers due to its excellent solubility in water. However, the dried PEDOT:PSS film is easy to form hard particles inside, which may induce fissure and then decrease electrical conductivity. It is not suitable for continuous bending and stretching. To solve this problem, porous substrates have been developed for printing and permeating PEDOT:PSS ink, such as fabrics and cellulose paper, which can greatly promote their adhesion. This strategy greatly improves the stability of wearable electromechanical sensor fabricated with PEDOT:PSS ink [30]. The polyvinyledenedifluoride (PVDF) is another appealing sensing material with many attractive properties, such as piezoelectric property, especially appropriate for piezoelectric wearable electromechanical sensors. Moreover, other conductive polymers such as PPy, poly(3-hexylthiophene-2,5-diyl) (P3HT) and PANI have also been utilized to fabricate wearable sensors [31]. More recently, ionic liquid (IL), a kind of salt that keep liquid state at room temperature, has attracted extensive attention [32]. Similar to liquid metals, IL can also be embedded in PDMS-based microchannels to fabricate wearable electromechanical sensor.

3.3 Performance of wearable electromechanical sensor

3.3.1 Wearable strain sensor

Wearable strain sensor converts strain into electrical signal. Many applications, such as human health monitoring, require enough stretchability range from tiny deformation (small than 1%) to large deformations (as large as 100%) and high sensitivity. There are two main strategies to enhance the sensitivity. One is choosing proper sensing materials. Various kinds of nanomaterials are tested, as seen in **Table 1**. For example, by coating graphene on woven fabric structure, a maximum elongation of 57% and a GF of 416 and 3667 at lower and higher strains are achieved. Combining graphene and nanocellulose into nanocomposite, it shows ultrahigh sensitivity with GF of 502 at 1% strain and 2427 at 6% strain.

The second strategy is structure engineering. As discussed in above section, cracks can greatly enhance the change of resistance. Network cracks formed in multilayer CNT films on PDMS composite result in both high gauge factor (maximum value of 87) and a wide sensing range (up to 100%) of the strain sensor, which allows the detection of strain as low as 0.007% with excellent stability (1500 cycles) [27].

To improve stretchability, many strategies have been developed. One strategy is using intrinsically flexible materials and the relative stiff components bridged with highly flexible interconnects [48]. When the intrinsic stretchability of flexible material is not enough, structural engineering can be used to further enhance their stretchability. The fragmented structure with connected islands can form a lot of cracks, which can relieve most of the applied strain through opening and enlargement

Material	Type	Sensitivity	Stretchability	Linearity	Durability (cycles)	Refs
AgNW	Strain	150,000	60%	0.989	200	[33]
AgNW	Pressure	1.54 kPa^{-1}	0.6 Pa-115 kPa	linear	5000	[34]
AuNW (gold nanowire)	Strain	70	250%	Nonlinear	500	[35]
AuNW	Pressure	1.14 kPa^{-1}	13 Pa-5 kPa	Linear	5000	[36]
Carbon black	Strain	647	20%	Nonlinear	200	[37]
Carbon black	Pressure	4.2 kPa^{-1}	0–30 kPa	0.996	30,000	[38]
Carbon nanofiber	Strain	72	300%	Nonlinear	8000	[39]
Carbon nanofiber	Pressure	4.2 kPa^{-1}	1.0 Pa-2 kPa	Nonlinear	10,000	[40]
Carbonized silk	Strain	9.6 (0–250%) 37.5 (250–500%)	500%	Nonlinear	10,000	[41]
Carbon nanotube	Strain	80	100%	Nonlinear	1500	[42]
Carbon nanotube	Pressure	0.209 kPa^{-1}	5.0 Pa-50 kPa	Nonlinear	5000	[43]
Graphene	Pressure	1.2 kPa^{-1}	0–25 kPa	Linear	1000	[44]
Graphene	Strain	1054	26%	Nonlinear	500	[45]
Mxene	Strain	64.6 (0–30%) 772.6 (30–70%)	130%	Nonlinear	5000	[46]
Mxene	Pressure	4.05 kPa^{-1} (0–1.0 kPa) 22.56 kPa^{-1} (1–3.5 kPa)	0–3.5 kPa	Nonlinear	10,000	[47]

Table 1.
Performance of wearable electromechanical sensor fabricated with typical nanomaterials.

of cracks. Deformable structures are widely used. For instance, the horseshoe and filamentary serpentine have been patterned with nanomaterials, which can accommodate large strain [49, 50]. Porous structures such as sponge and foam are also employed to improve the stretchability [51]. Wrinkled structure based on CNT film is produced and integrated on an Ecoflex substrate, allowing conductivity up to 750% elongation, an approximate 60 times increase versus nonwrinkled films [52].

Significant progress has been achieved on the sensitivity and stretchability, but there are some challenges still existing. Most resistive wearable strain sensors suffer from at least one of these problems, which are nonlinear response, large hysteresis, and irreversibility. The irreversibility mainly origins from partial slides back of sensing materials and irreversibly recover of cracks. Hysteresis is mainly caused by the viscoelasticity of polymers and the friction between the sensing materials and the polymer matrix. The rearrangement of sensing materials and opening of cracks are also responsible for time delay between electrical output and mechanical input. Nonlinear response mainly results from crack propagation and tunneling effect, which is always exponential as discussed above. Therefore, the performance of resistive wearable strain sensor should be evaluated from more aspects in further research.

Compared with resistive wearable strain sensor, capacitive strain sensors possess good linearity with low hysteresis, fast response, and are less susceptible to overshoot and creep. Nanomaterial-based stretchable conductors are usually used as the electrodes for capacitive strain sensors. Highly stretchable silicone, such as PDMS, Dragon Skin, and Ecoflex are commonly used as the dielectric layer sandwiched between two electrodes. For example, a capacitive strain sensor is fabricated with stretchable AgNW/PDMS conductors as the top and bottom electrodes and Ecoflex as the dielectric material [53]. The GF of this sensor reaches 0.7 and its stretchability is up to 50%. Moreover, it also has a good linearity. While capacitive strain sensors exhibit smaller GFs than the resistive strain sensors, they are ideal for applications where the strain is relatively large. In addition, the GFs of capacitive strain sensors remain constant in the entire strain range.

3.3.2 Wearable pressure sensor

Wearable pressure sensor converts pressure into electrical signal. Pressure sensor can be fabricated with interlocked structures, percolative networks of nanomaterials, microfabricated structures (e.g., micropyramids, micropillars), porous structures (e.g., sponges, foams, porous rubbers), and so forth. For example, **Figure 3a** presents a pressure sensor fabricated with interlocked microdome array. The contact between microdome increases when pressure is applied, thus decreasing the tunneling resistance [54].

To improve the sensitivity of piezoresistive pressure sensor, structural surface modification of the electrodes is an effective strategy. Incorporation of nano/microscaled structures can provide large changes in contact resistance, allowing for detections of smaller pressures. For example, through coating polyurethane sponge with graphene to form fracture structure, a two-order of magnitude increase in sensitivity within the 0–2 kPa regime is demonstrated compared with no fracture one [55].

For the capacitive pressure sensor, the separation between two electrodes decreases with the pressure, resulting in an increase in capacitance. The property of dielectric materials almost determines the pressure sensitivity. Lower elastic modulus means a larger strain ε under a given pressure. The dielectric constant increased with pressure and low Poisson's ratio would all benefit the performance. High sensitivity of 0.8 kPa^{-1} has been reported by using a GO-based low elastic modulus foam as the dielectric material [56]. There are several methods been demonstrated to fabricate highly deformable dielectric materials, including using commercial

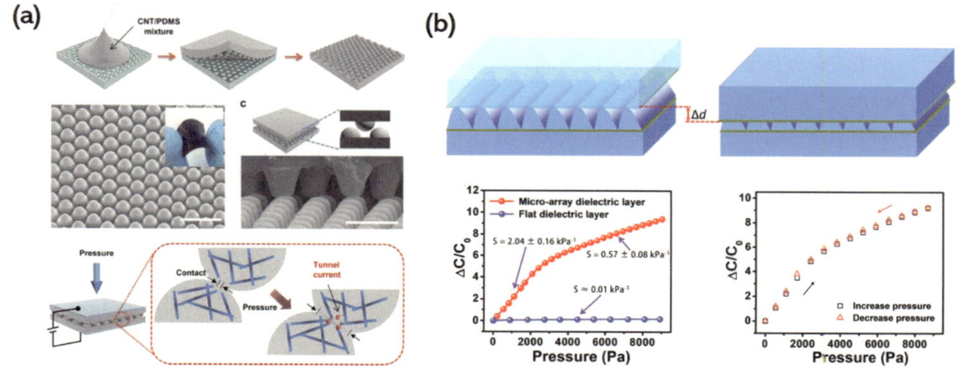

Figure 3.
(a) Schematic of the fabrication procedure and mechanism of pressure with interlocked microdome arrays. (b) Response characteristics of the flexible capacitive pressure sensor based on the PDMS microarray dielectric layer.

porous tapes, using special molds (e.g., the surface of matte glass, a micromachined Si mold, or the surface of lotus leaf) to create microstructures in elastomers, using sugar cubes as the template to create porous elastomers and fabricating buckled structures through prestretching and releasing. As the dielectric constant of air is smaller than that of the dielectric material used for the sensor, the effective dielectric constant is increased under pressure when the air gap is compressed. For example, **Figure 3b** shows a flexible pressure sensor with high sensitivity been built, which is a typical sandwich structure by combining a microarrayed PDMS dielectric layer with PDMS substrates. The top/bottom electrode material is PDMS substrate coated with AgNWs, and the dielectric layer is a PDMS with microarray structure, which is used to improve the pressure sensitivity. The results show that it possesses high sensitivity (2.04 kPa^{-1}) in low-pressure ranges (0–2000 Pa), low detection limits (<7 Pa), and fast response times (<100 ms). Meanwhile, it also has excellent bending and cycling stability [57].

Progress has also been made on wearable piezoelectric and triboelectric pressure sensors. For example, it has been reported that a novel piezoelectric pressure sensor was fabricated through sandwiching freestanding electrospun polyvinyledenedifluoride-trifluoroethylene (PVDF-TrFE) nanofiber arrays [58] or electrospun PVDF-TrFE nanofiber between two electrodes. It can detect very tiny pressures as low as 0.1 Pa and has high sensitivity up to 1.1 V kPa^{-1} for pressure range from 0.4–2 kPa. In a representative work, a pressure-responsive triboelectric nanogenerator is used to gate the graphene transistors. Such graphene tribotronics showed a pressure sensitivity of $\approx$2% kPa^{-1} at a pressure of 10 kPa.

4. Fabrication technology of wearable electromechanical sensor

The wearable electromechanical sensor usually consists of three basic components, which are substrate, active elements, and electrode/interconnect. They are usually fabricated with different materials. During the fabrication process, combining the substrate and active elements is the key step. Basically, there are two situations. One is that sensing material forms uniform composite with polymer substrate, the other is that sensing material is attached on substrate and a clear interface exists. In this part, we will focus on the combination strategies for substrates and sensing elements, and some key processes for performance enhancement are also concerned.

4.1 Fabrication of wearable composite electromechanical sensor

For the composite electromechanical sensor, the substrate and sensing materials should be fabricated into composite. The key process is how to mix them and prepare uniform composite. The sensing materials are usually mixed with polymers by magnetically or ultrasonically stirring, and then the dried elastic composites can be prepared in bulk or film forms. The mixed composites have complex electromechanical features that are induced by the diversity of sensing materials and polymer and significantly depend on concentration of sensing materials and its distribution state. For example, the electrical property of carbon black-silicone composite is mainly determined by carbon black concentration. The electrical resistance clearly increases with the applied uniaxial pressure when the concentration is about 0.08–0.09 wt%. By further increasing the concentration from 0.1 to 0.13 wt%, the change tendency of electrical resistance switches from increase to decrease. Finally, the electrical resistance starts to decrease with the uniaxial pressure with the concentration larger than 0.14 wt % [59].

4.2 Fabrication of wearable layered electromechanical sensor

For the wearable layer electromechanical sensor, the substrate and sensing materials are assembled into film layer by layer. Many techniques have been developed to assemble active material on substrate, including printing, coating, casting, and other methods.

Printing can simultaneously deposit and pattern many materials on various substrates without the need for sophisticated equipment and clean room. The wearable sensors can be printed with/without the help of masks, according to the specific implementation approach, as seen in **Figure 4a** [60]. The electrode pattern can directly be obtained by inkjet printing. Inkjet printing is an accurate, fast, and reproducible film preparation technique. Functional ink droplets are propelled onto different substrates by a nozzle. The functional inks should have proper solubility, viscosity, and surface tension. As a typical printing method, screen printing requires the help of mask and proper functional ink. During the process, screen openings are fully covered with functional by using fill blade or squeegee, and then it is transferred onto substrate surface. Finally, the mask is removed, and a patterned film is formed on the substrate by functional ink. This technique has been widely used in manufacturing sensing materials in electromechanical sensors.

Lithography is a pattern transferring method to realize diverse and ingenious geometries. This process firstly deposits functional layer onto the substrate and then etches the undesired areas by reagent solutions with the help of photolithography. Since photolithography and wet etching has high accuracy, the devices with sophisticated geometries and rich functionality can be obtained. Coating technique is another popular method because of its low cost and simplicity. There are different advantages for different coating methods. Dip coating can be used to any kinds of substrate and can control the thickness by dipping time. Spin coating is easy to form uniform film and can control the thickness by time and spin speed. Compared with spin and dip coating, spray coating can fully utilize the functional inks. **Figure 4b** shows a buckled sheath-core fiber-based ultrastretchable sensor fabricated with spray coating methods. The fiber wearable strain sensor possesses excellent stretchability higher than 1135% and fast response time ($\approx$16 ms). Moreover, the performance is very repeatable and stable even after 20,000 cycles with loading/unloading test [47].

Novel techniques have been developed, such as laser scribed (LS) technique. Graphene oxide (GO) can be simultaneously reduced and patterned by laser [61]. Carbonating substrate material by one-step direct laser writing (DLW) has also been validated. Glassy and porous carbon structures have been produced from PI film via DLW. The DLW-based graphene possesses favorable electroconductibility, porousness, and superhydrophilic wettability. Directly drawing electronics with various instruments has recently become an alternative technique. This technique endows end-users the capability to design and realize sensors according to the "on-site, real-time" demands [62]. "Penciling it on" has been proved to be a simple, rapid, and solvent-free method for producing electronics [63]. Chinese brush pen is a possible more appealing writing instrument for sensor fabrication. Similarly, the animal hair bundle is first soaked into low-viscosity ink, and then the ink is uniformly coated on the substrate by well-controlled handwriting manner. Benefiting from excellent liquid manipulation of Chinese brush pen, sensing materials can be coated on different substrates without considering its rigidness and surface roughness. For example, a high-performance tattoo-like strain sensor has been fabricated with AuNWs/PANI ink writing by Chinese brush pen [64]. Various types of functional inks can be loaded in their reservoirs, including metal inks, liquid metals, and even organic mixtures. Sophisticated structures can be generated with controllable geometries on many substrates by using these two methods [65]. Wet spinning is

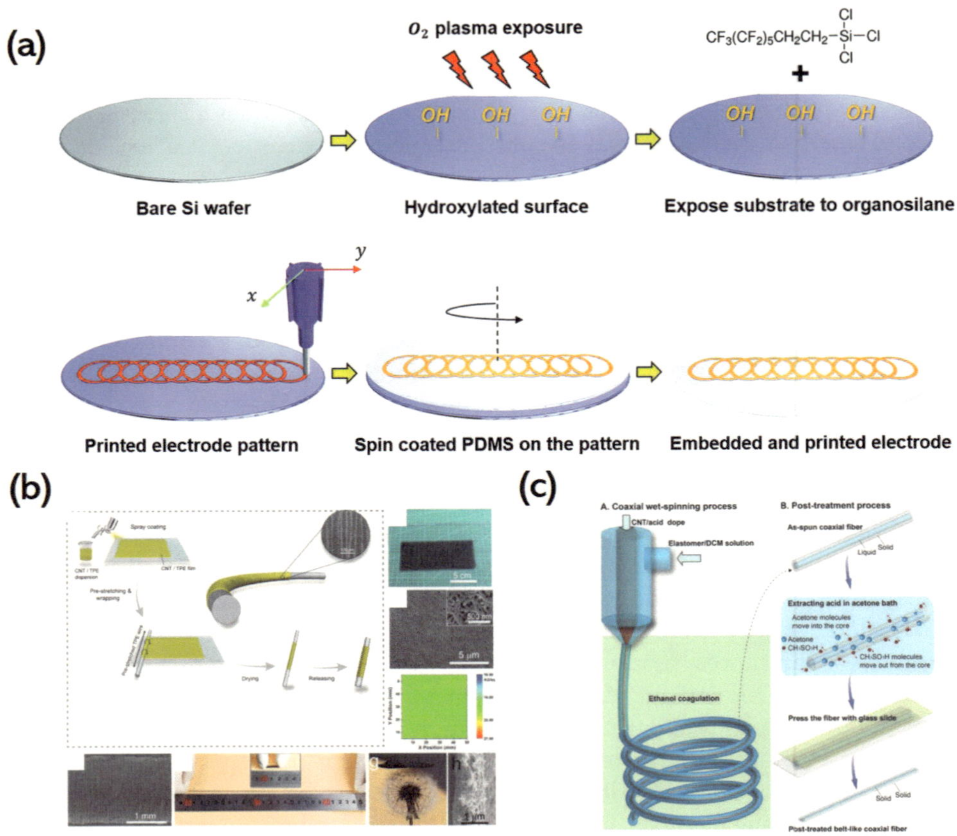

Figure 4.
Wearable electromechanical sensor fabricated with different techniques: (a) inkjet printing, (b) dropping casting, and (c) spray coating.

another special method to fabricate fiber shape wearable electromechanical sensor. **Figure 5c** shows a fiber strain sensor fabricated with coaxial wet-spinning and post-treatment process. The spinning nozzle has the coaxial inner and outer channels, respectively. The inner spinning dope is SWCNT/CH_3SO_3H, and the outer spinning solution is the solution of thermoplastic elastomer (TPE) in CH_2Cl_2. The SWCNT/CH_3SO_3H dope from the inner channel and the TPE/CH_2Cl_2 solution from the outer channel are introduced into the ethanol coagulation bath simultaneously. A single TPE-wrapped SWCNT coaxial fiber is then wetspun and collected successfully. The sensors attain high sensitivity (with a gauge factor of 425 at 100% strain), high stretchability, and high linearity.

4.3 Fabrication of wearable 3D electromechanical sensor

For the wearable 3D electromechanical sensor, the substrate and sensing materials are combined into 3D structure. The first method introduced is microscale modeling. It is often utilized to fabricate different microstructures in substrates, electrodes, and sensing composites. Successfully designed microstructure not only can be used to increase the sensitivity of piezoresistive but also that of capacitive sensors when microstructured dielectric is applied. Different modules have been developed, including micromachined wafers, silk fabrics, and even plant leaves. During the fabrication process, sensing materials are simply poured onto the module and peeled off after partial or complete drying. The adhesion between processed

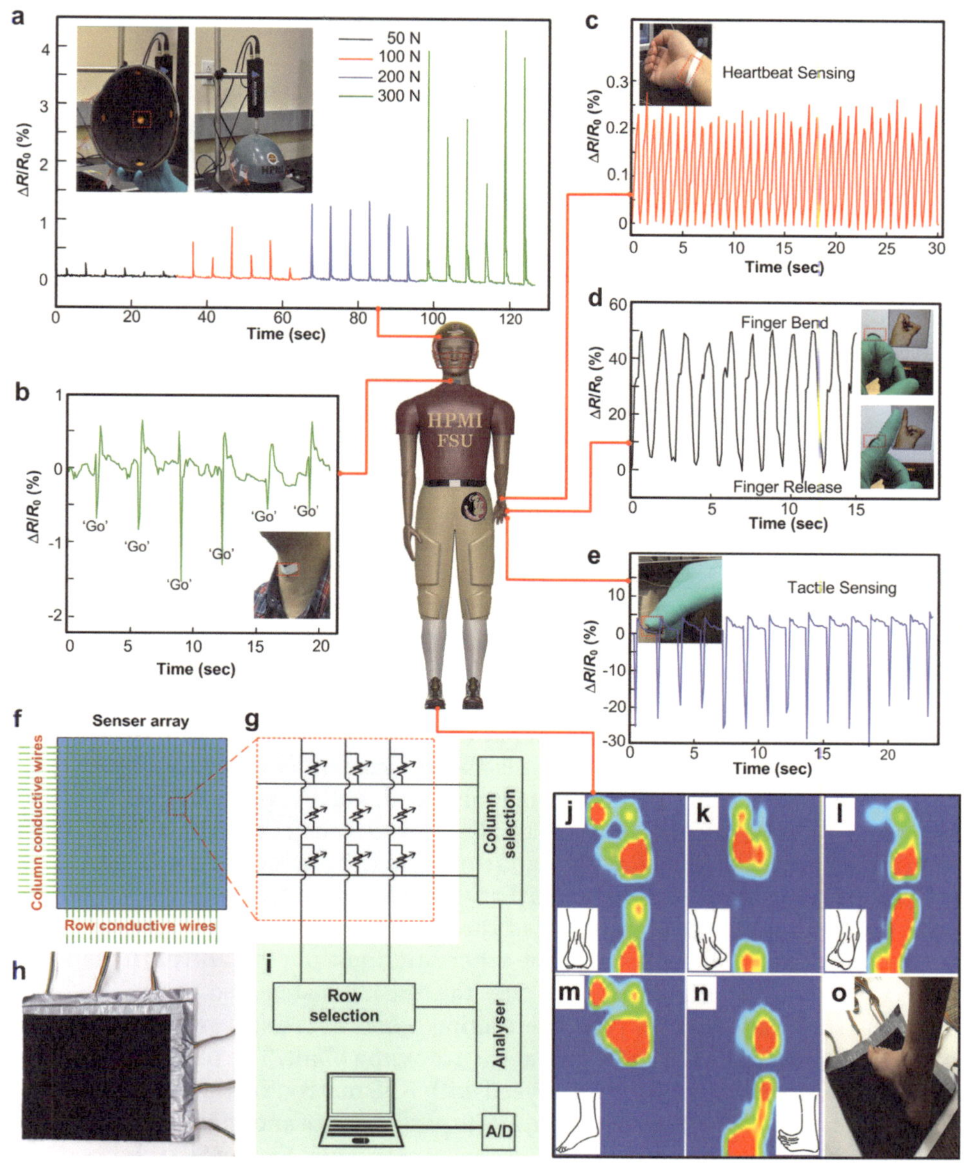

Figure 5.
Health motion monitoring with CNT-based strain sensor: (a) impact pressure, (b) muscle movement, (c) heartbeat, (d) finger motion, (e) finger touch, (f) schematic diagram of sensor array, (g) magnified view of the sensor array, (h) optical photograph of a fabricated sensor array containing 25 × 25 pixels, (i) circuit schematic of the sensor matrix (j–n) foot pressure pattern, and (o) strain sensor attached on the human right foot.

material and module is the most important parameters for this technique, which can be adjusted by necessary pretreatment and sophisticated geometric design.

3D printing is the best candidate for developing 3D constructions and has gained great popularity due to its powerful ability [66]. If the sensing materials are well prepared, arbitrary structures can be printed with 3D printing with adjustable resolution, even lower than 0.1 μm. For instance, A three-layer sensor has been fabricated in a single step by 3D printing, which originally requires multiple steps by using traditional method, including micromolding, laminating, and infilling. Wearable pressure sensor has also been realized by a multimaterial, multiscale, and multifunctional 3D printing approach. The size of this sensor is 3 × 3 mm in area and 1.2 mm in height [67].

5. Applications of wearable electromechanical sensor

Wearable electromechanical sensors can basically detect mechanical signals including pressure and strain. Applications that require monitoring pressure and strain are theoretically can realized by it. Until now, monitoring of human motion and health, speech recognition, gesture recognition, human machine interaction, acoustic waves detection, and even disease diagnosis have been demonstrated, which would be discussed in the following.

5.1 Human motion monitoring

When wearable electromechanical sensor is mounted on the skin or integrated with textiles, it can real-time monitor human motions including hand, limb, foot, face, and throat. Subtle deformations induced by body activities such as blood pulse flow and respiration, and large deformations related to the body movements such as finger and knee bending can be readily detected. **Figure 5** shows the human motions in daily life detected by CNT-coated auxetic foam strain sensor (AFS) [68]. As **Figure 5a** shows, the foam sensors performed well by dependably detecting the timing, frequency, and magnitude of the impact event and outputting signals in sharp spikes corresponding to the impact events. **Figure 5b** shows the monitoring of the muscle movement during speech by attaching a foam sensor onto a person's neck. When the person repeatedly says the simple words "go," stable signals can observe which timing and pattern corresponded well with the vocal events. Moreover, the wrist pulse has also been successfully monitored by the AFS (**Figure 5c**). A typical pulse waveform is obtained, and the pulse frequency of 76 beats min^{-1} can be calculated. It can also be used to transfer the human intentions of pressing buttons and switches by attaching the AFS directly to the fingertip (**Figure 5d**). **Figure 5e** demonstrates that the AFS can control gesture by wearing on the finger joint because the signal of the foam sensor one by one corresponds to the gesture. **Figure 5f** and **g** shows the schematic and a photograph of the sensor matrix, respectively. **Figure 5h** illustrates the sensing system and a simplified electrical schematic that scan the intersecting points of the sensor's rows and columns and measure the resistance at each crossing point. The plantar pressure distribution can be successfully analyzed with AFS matrix, further extending its fields of application ranging from sports performance and injury prevention to prosthetics and orthotics design. For further example, **Figure 5j–n** shows the various barefoot pressure distributions applied by a human right foot (**Figure 5o**), including neutral position, pronation, supination, plantar flexion, and dorsiflexion, which is displayed by the colored contour maps. The in-shoe plantar pressure measurement can also be finished by simply inserting AFS matrix into shoes. It can be anticipated that wearable electromechanical sensor can find a wide range of applications in human motion monitoring, body pressure distribution, and even adjusting sitting posture.

5.2 Human health monitoring

Human health monitoring is based on the continuous monitoring of human motions, especially the pulse and respiration. Wearable electromechanical sensor attached on wrist and chest can be used to detect the pulse and respiratory rate. **Figure 6** shows that graphene film strain sensor can exactly monitor people's pulse and breath rate. Strain sensor are attached on a person's wrist or chest for real-time recording of pulse and respiratory rate signals (**Figure 7a**) [69]. **Figure 6b** shows the collected pulse and respiratory signals, where each cycle represents a pulse or

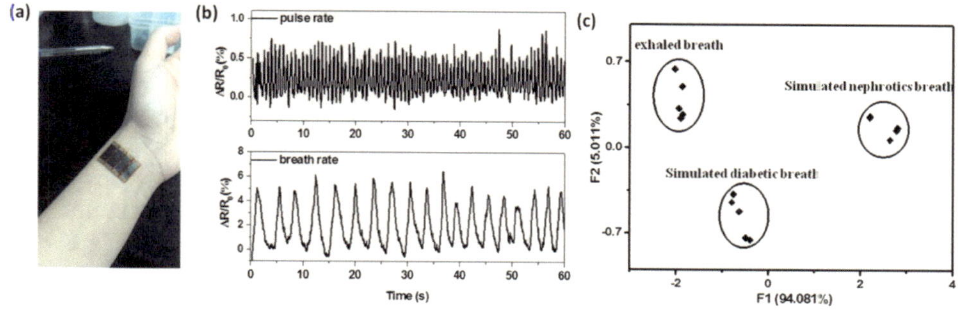

Figure 6.
Health monitoring with graphene strain sensor. (a) Photograph of strain sensor mounted on the human wrist, (b) normalized resistance changes of the strain sensor when monitoring wrist pulses and respiratory rate, and (c) PCA analysis of exhaled breath of simulated nephrotic patients, diabetic patients, and healthy people.

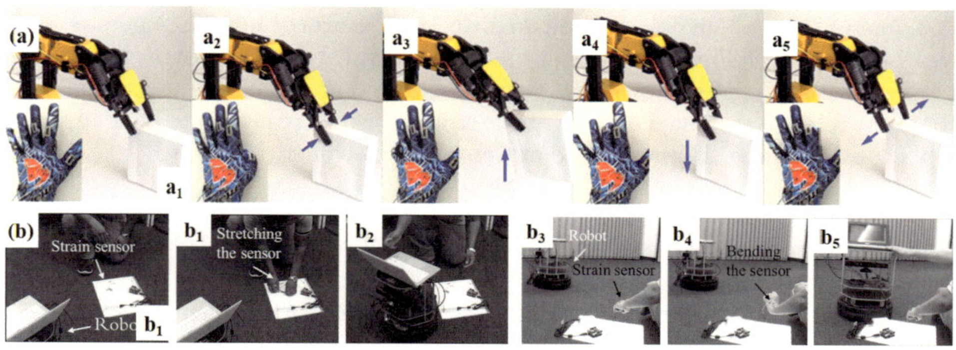

Figure 7.
Piezoresistive sensors for human-machine interfaces: (a) smart gloves and (b) robotic controlling.

breath. The valleys correspond to the shrinking of the chest, and peaks represent the stretching of the chest. Then, the pulse and breath rates can be estimated to be about 76 and 19 in 60 s, respectively. Three kinds of exhaled breath (simulated diabetic breath, simulated nephrotic breath, and the breath of healthy individuals) are investigated. The obtained response data are analyzed, and the results are displayed in **Figure 6c**. It can be observed that the three breath samples are clearly different. The exhaled breath samples are categorized into three distinguishable clusters without any overlap, which correspond to healthy individuals, simulated diabetic patients, and simulated nephrotic patients, respectively. This demonstrates that wearable strain sensor has high potential for human health monitoring and even the diagnosis diseases.

5.3 Speech recognition

Speech recognition is also based on the monitoring of human motions. When the wearable electromechanical is attached on the throat, it could record muscular movements in order to collect and recognize speech sounds. This is permitted by the fact that the throat muscle exhibits different degrees of stretching or shrinking strains when speaking different words. Due to the tiny changes caused by throat motion, the strain sensor used in speech recognition should have high sensitivity. The GF of GWF strain sensor can be as high as 10^3 with 2–6% strains, 106 with higher strains (>7%), and ~35 with a minimal strain of 0.2%, which is suitable for this application. The results show test signal waveforms of all 26 english letters [70]. As expected, the waveforms are unique and repeatable for all letters. Since each individual speech organ is different, people can easily distinguish whether a given voice comes from the same person. This demonstrates that wearable electromechanical can be used in speech recognition.

5.4 Human-machine interface

Human-machine interfaces and robotic remote controlling are greatly beneficial in surgery or a highly risky work that requires the replacement of robotics. The electromechanical sensor used in human-machine interface is typically mounted on body joint, which are normally bended or stretched at large degree of deformations; thus, high stretchability (>50%) is required. The robotic controlling is demonstrated in **Figure 7**, and the wearable strain sensors are based on the hybrid of polyaniline and gold nanowires for a smart glove [9]. The sensor-based-smart glove is used to control the movement of a robot through wireless signals (**Figure 7a**). The robot is at relaxed state (a_1) and works as an arm that can clamp (a_2), lift up (a_3), put down (a_4), and release (a_5) an object based on different postures of human fingers as wearing the sensor. **Figure 7b** reveals the remote control on the robot movement by a strain sensor based on graphene. As can be seen in this figure, b_1 and b_4 demonstrate the robot at the relaxed state. As the strain sensor is stretched or bended, the robot starts working (b_2 and b_5) and moves to the controller (b_3 and b_6).

6. Conclusion and outlook

In this chapter, we discuss the working mechanism, fabrication methods, and applications of wearable electromechanical sensors. Piezoresistive sensor attracts more attentions due to its clear structure, mechanism, fabrication methods, and low cost. High sensitivity and stretchability have been achieved simultaneously. However, the stability and linearity are still limited for resistive-type sensor. Moreover, mass production with low cost is still a challenge. One strategy to reduce cost is developing novel fabrication methods, which can readily build high-performance sensor. Many applications have been demonstrated in a qualitative way by using strain or pressure sensor. However, the practical application needs more quantitative analysis, which requires further investigations.

Acknowledgements

This work is supported by the Start-up Research Grant (SRG2016-00092-IAPME), Multi-Year Research Grant (MYRG2018-00079-IAPME) of University of Macau, Science and Technology Development Fund (081/2017/A2), (0059/2018/A2), (009/2017/AMJ), Macao SAR (FDCT).

Conflict of interest

There is no conflict of interest between authors.

DOI: http://dx.doi.org/10.5772/intechopen.85098

Author details

Dan Liu and Guo Hong*
Institute of Applied Physics and Materials Engineering, University of Macau, Taipa, Macau

*Address all correspondence to: ghong@um.edu.mo

References

[1] Abshirini M, Charara M, Liu Y, Saha M, Altan MC. 3D printing of highly stretchable strain sensors based on carbon nanotube nanocomposites. Advanced Engineering Materials. 2018;**20**:1800425

[2] Jung YH, Park B, Kim JU, Kim TI. Bioinspired electronics for artificial sensory systems. Advanced Materials. 2018:e1803637

[3] Sun Y, Yang L, Xia K, Liu H, Han D, Zhang Y, et al. "Snowing" graphene using microwave ovens. Advanced materials. 2018;**30**:e1803189

[4] Park M, Bok BG, Ahn JH, Kim MS. Recent advances in tactile sensing technology. Micromachines (Basel). 2018;**9**:321

[5] Schroeder V, Savagatrup S, He M, Lin S, Swager TM. Carbon nanotube chemical sensors. Chemical Reviews. 2019;**119**:599

[6] Maity D, Kumar RTR. Polyaniline anchored MWCNTs on fabric for high performance wearable ammonia sensor. ACS Sens. 2018;**3**:1822

[7] Amjadi M, Kyung K-U, Park I, Sitti M. Stretchable, skin - mountable, and wearable strain sensors and their potential applications: A review. Advanced Functional Materials. 2016;**26**:1678

[8] Yang T, Xie D, Li Z, Zhu H. Recent advances in wearable tactile sensors: Materials, sensing mechanisms, and device performance. Materials Science and Engineering. 2017;**115**:1

[9] Jason NN, Ho MD, Cheng W. Resistive electronic skin. Journal of Materials Chemistry C. 2017;**5**:5845

[10] Tolvanen J, Hannu J, Jantunen H. Stretchable and washable strain sensor based on cracking structure for human motion monitoring. Scientific Reports. 2018;**8**:13241

[11] Heikenfeld J, Jajack A, Rogers J, Gutruf P, Tian L, Pan T, et al. Wearable sensors: Modalities, challenges, and prospects. Lab on a Chip. 2018;**18**:217

[12] Yin B, Liu X, Gao H, Fu T, Yao J. Bioinspired and bristled microparticles for ultrasensitive pressure and strain sensors. Nature Communications. 2018;**9**:5161

[13] Minot ED, Yaish Y, Sazonova V, Park J-Y, Brink M, McEuen PL. Tuning carbon nanotube band gaps with strain. Physical Review Letters. 2003;**90**:156401

[14] Amjadi M, Pichitpajongkit A, Lee S, Ryu S, Park I. Highly stretchable and sensitive strain sensor based on silver nanowire–elastomer nanocomposite. ACS Nano. 2014;**8**:5154

[15] Hong SK, Yang S, Cho SJ, Jeon H, Lim G. Development of a waterproof crack-based stretchable strain sensor based on PDMS shielding. Sensors. 2018;**18**:1171

[16] Yang T, Wang W, Zhang H, Li X, Shi J, He Y, et al. Tactile sensing system based on arrays of graphene woven microfabrics: Electromechanical behavior and electronic skin application. ACS Nano. 2015;**9**:10867

[17] Cai G, Wang J, Qian K, Chen J, Li S, Lee PS. Extremely stretchable strain sensors based on conductive self-healing dynamic cross-links hydrogels for human-motion detection. Advanced Science. 2017;**4**:1600190

[18] Liu X, Su G, Guo Q, Lu C, Zhou T, Zhou C, et al. Hierarchically structured self-healing sensors with tunable positive/negative piezoresistivity. Advanced Functional Materials. 2018;**28**:1706658

[19] Poland CA, Duffin R, Kinloch I, Maynard A, Wallace WAH, Seaton A, et al. Carbon nanotubes introduced into the abdominal cavity of mice show asbestos-like pathogenicity in a pilot study. Nature Nanotechnology. 2008;**3**:423

[20] Wang C, Xia K, Jian M, Wang H, Zhang M, Zhang Y. Carbonized silk georgette as an ultrasensitive wearable strain sensor for full-range human activity monitoring. Journal of Materials Chemistry C. 2017;**5**:7604

[21] Kuang SY, Chen J, Cheng XB, Zhu G, Wang ZL. Two-dimensional rotary triboelectric nanogenerator as a portable and wearable power source for electronics. Nano Energy. 2015;**17**:10

[22] Vyas R, Lakafosis V, Lee H, Shaker G, Yang L, Orecchini G, et al. Inkjet printed, self powered, wireless sensors for environmental, gas, and authentication-based sensing. IEEE Sensors Journal. 2011;**11**:3139

[23] Yiin-Kuen F, Hsi-Chun H. Highly flexible self-powered sensors based on printed circuit board technology for human motion detection and gesture recognition. Nanotechnology. 2016;**27**:095401

[24] Hudak NS, Amatucci GG. Small-scale energy harvesting through thermoelectric, vibration, and radiofrequency power conversion. Journal of Applied Physics. 2008;**103**:101301

[25] Zhong J, Zhong Q, Hu Q, Wu N, Li W, Wang B, et al. Stretchable self-powered fiber-based strain sensor. Advanced Functional Materials. 2015;**25**:1798

[26] Taroni PJ, Santagiuliana G, Wan K, Calado P, Qiu M, Zhang H, et al. Toward stretchable self-powered sensors based on the thermoelectric response of PEDOT:PSS/polyurethane blends. Advanced Functional Materials. 2017;**28**:1704285

[27] Lin CW, Zhao Z, Kim J, Huang J. Pencil drawn strain gauges and chemiresistors on paper. Scientific Reports. 2014;**4**:3812

[28] Nag A, Mitra A, Mukhopadhyay SC. Graphene and its sensor-based applications: A review. Sensors and Actuators A: Physical. 2018;**270**:177

[29] Yin B, Wen Y, Hong T, Xie Z, Yuan G, Ji Q, et al. Highly stretchable, ultrasensitive, and wearable strain sensors based on facilely prepared reduced graphene oxide woven fabrics in an ethanol flame. ACS Applied Materials & Interfaces. 2017;**9**:32054

[30] Eom J, Jaisutti R, Lee H, Lee W, Heo JS, Lee JY, et al. Highly sensitive textile strain sensors and wireless user-interface devices using all-polymeric conducting fibers. ACS Applied Materials & Interfaces. 2017;**9**:10190

[31] Jia HY, Wang J, Zhang XY, Wang YP. Pen-writing polypyrrole arrays on paper for versatile cheap sensors. ACS Macro Letters. 2014;**3**:86

[32] Yoon SG, Koo HJ, Chang ST. Highly stretchable and transparent microfluidic strain sensors for monitoring human body motions. ACS Applied Materials & Interfaces. 2015;**7**:27562

[33] Liao X, Zhang Z, Kang Z, Gao F, Liao Q, Zhang Y. Ultrasensitive and stretchable resistive strain sensors designed for wearable electronics. Materials Horizons. 2017;**4**:502

[34] Wan Y, Qiu Z, Huang J, Yang J, Wang Q, Lu P, et al. Natural plant materials as dielectric layer for highly sensitive flexible electronic skin. Small. 2018;**14**:e1801657

[35] Boland CS, Khan U, Benameur H, Coleman JN. Surface coatings of silver nanowires lead to effective, high conductivity, high-strain, ultrathin sensors. Nanoscale. 2017;**9**:18507

[36] Gong S, Schwalb W, Wang Y, Chen Y, Tang Y, Si J, et al. A wearable and highly sensitive pressure sensor with ultrathin gold nanowires. Nature Communications. 2014;**5**:3132

[37] Song H, Zhang J, Chen D, Wang K, Niu S, Han Z, et al. Superfast and high-sensitivity printable strain sensors with bioinspired micron-scale cracks. Nanoscale. 2017;**9**:1166

[38] Fan X, Huang Y, Ding X, Luo N, Li C, Zhao N, et al. Alignment - free liquid - capsule pressure sensor for cardiovascular monitoring. Advanced Functional Materials. 2018;**28**:1805045

[39] Ding Y, Yang J, Tolle CR, Zhu Z. A highly stretchable strain sensor based on electrospun carbon nanofibers for human motion monitoring. RSC Advances. 2016;**6**:79114

[40] Peng S, Blanloeuil P, Wu S, Wang CH. Rational design of ultrasensitive pressure sensors by tailoring microscopic features. Advanced Materials Interfaces. 2018;**5**:1800403

[41] Wang C, Li X, Gao E, Jian M, Xia K, Wang Q, et al. Carbonized silk fabric for ultrastretchable, highly sensitive, and wearable strain sensors. Advanced Materials. 2016;**28**:6640

[42] Wang S, Xiao P, Liang Y, Zhang J, Huang Y, Wu S, et al. Network cracks-based wearable strain sensors for subtle and large strain detection of human motions. Journal of Materials Chemistry C. 2018;**6**:5140

[43] Gao Y, Yu G, Tan J, Xuan F. Sandpaper-molded wearable pressure sensor for electronic skins. Sensors and Actuators A: Physical. 2018;**280**:205

[44] Shi J, Wang L, Dai Z, Zhao L, Du M, Li H, et al. Multiscale hierarchical design of a flexible piezoresistive pressure sensor with high sensitivity and wide linearity range. Small. 2018;**14**:e1800819

[45] Yang YF, Tao LQ, Pang Y, Tian H, Ju ZY, Wu XM, et al. An ultrasensitive strain sensor with a wide strain range based on graphene armour scales. Nanoscale. 2018;**10**:11524

[46] Cai Y, Shen J, Ge G, Zhang Y, Jin W, Huang W, et al. Stretchable Ti_3C_2Tx MXene/carbon nanotube composite based strain sensor with ultrahigh sensitivity and tunable sensing range. ACS Nano. 2018;**12**:56

[47] Ma Y, Yue Y, Zhang H, Cheng F, Zhao W, Rao J, et al. 3D Synergistical MXene/reduced graphene oxide aerogel for a piezoresistive sensor. ACS Nano. 2018;**12**:3209

[48] Liu H, Gao J, Huang W, Dai K, Zheng G, Liu C, et al. Electrically conductive strain sensing polyurethane nanocomposites with synergistic carbon nanotubes and graphene bifillers. Nanoscale. 2016;**8**:12977

[49] Jason NN, Wang SJ, Bhanushali S, Cheng W. Skin inspired fractal strain sensors using a copper nanowire and graphite microflake hybrid conductive network. Nanoscale. 2016;**8**:16596

[50] Lim S, Son D, Kim J, Lee YB, Song J-K, Choi S, et al. Transparent and stretchable interactive human machine interface based on patterned graphene heterostructures. Advanced Functional Materials. 2015;**25**:375

[51] Zhang R, Hu R, Li X, Zhen Z, Xu Z, Li N, et al. A bubble-derived strategy to prepare multiple graphene-based porous materials. Advanced Functional Materials. 2018;**28**:1705879

[52] Park S-J, Kim J, Chu M, Khine M. Highly flexible wrinkled carbon nanotube thin film strain sensor to monitor human movement. Advanced Materials Technologies. 2016;**1**:1600053

[53] Yao S, Zhu Y. Wearable multifunctional sensors using printed stretchable conductors made of silver nanowires. Nanoscale. 2014;**6**:2345

[54] Chun S, Choi IY, Son W, Bae GY, Lee EJ, Kwon H, et al. A highly sensitive force sensor with fast response based on interlocked arrays of indium tin oxide nanosprings toward human tactile perception. Advanced Functional Materials. 2018;**28**:1804132

[55] Yao H-B, Ge J, Wang C-F, Wang X, Hu W, Zheng Z-J, et al. A flexible and highly pressure - sensitive graphene–polyurethane sponge based on fractured microstructure design. Advanced Materials. 2013;**25**:6692

[56] Wan S, Bi H, Zhou Y, Xie X, Su S, Yin K, et al. Graphene oxide as high-performance dielectric materials for capacitive pressure sensors. Carbon. 2017;**114**:209

[57] Wang J, Jiu J, Nogi M, Sugahara T, Nagao S, Koga H, et al. A highly sensitive and flexible pressure sensor with electrodes and elastomeric interlayer containing silver nanowires. Nanoscale. 2015;**7**:2926

[58] Persano L, Dagdeviren C, Su Y, Zhang Y, Girardo S, Pisignano D, et al. High performance piezoelectric devices based on aligned arrays of nanofibers of poly(vinylidenefluoride-co-trifluoroethylene). Nature Communications. 2013;**4**:1633

[59] Kai W, Hirota Y, Hua L, Inoue Y. Thermal and mechanical properties of a poly(ϵ-caprolactone)/graphite oxide composite. Journal of Applied Polymer Science. 2008;**107**:1395

[60] Lee H, Lee J, Seong B, Jang H-S, Byun D. Printing conductive micro-web structures via capillary transport of elastomeric ink for highly stretchable strain sensors. Advanced Materials Technologies. 2018;**3**:1700228

[61] Tao LQ, Wang DY, Tian H, Ju ZY, Liu Y, Chen YQ, et al. Tunable and wearable high performance strain sensors based on laser patterned graphene flakes. 2016 IEEE Inter. Electron Devices Meeting. IEEE; 2016

[62] Li ZD, Liu H, Ouyang C, Wee WH, Cui XY, Lu TJ, et al. Recent advances in pen-based writing electronics and their emerging applications. Advanced Functional Materials. 2016;**26**:165

[63] Kanaparthi S, Badhulika S. Low cost, flexible and biodegradable touch sensor fabricated by solvent-free processing of graphite on cellulose paper. Sensors and Actuators B: Chemical. 2017;**242**:857

[64] Gong S, Lai DTH, Wang Y, Yap LW, Si KJ, Shi QQ, et al. Tattoolike polyaniline microparticle-doped gold nanowire patches as highly durable wearable sensors. ACS Applied Materials & Interfaces. 2015;**7**:19700

[65] Russo A, Ahn BY, Adams JJ, Duoss EB, Bernhard JT, Lewis JA. Pen-on-Paper flexible electronics. Advanced Materials. 2011;**23**:3426

[66] Leigh SJ, Bradley RJ, Purssell CP, Billson DR, Hutchins DA. A simple, low-cost conductive composite material for 3D printing of electronic sensors. PLoS One. 2012;**7**

[67] Guo SZ, Qiu KY, Meng FB, Park SH, McAlpine MC. 3D printed stretchable tactile sensors. Advanced Materials. 2017;**29**:1701218

[68] Li Y, Luo S, Yang M-C, Liang R, Zeng C. Poisson ratio and piezoresistive sensing: A new route to high-performance 3D flexible and stretchable sensors of multimodal sensing capability. Advanced Functional Materials. 2016;**26**:2900

[69] Xu H, Xiang JX, Lu YF, Zhang MK, Li JJ, Gao BB, et al. Multifunctional wearable sensing devices based

on functionalized graphene films for simultaneous monitoring of physiological signals and volatile organic compound biomarkers. ACS Applied Materials & Interfaces. 2018;**10**:11785

[70] Wang Y, Yang T, Lao J, Zhang R, Zhang Y, Zhu M, et al. Ultra-sensitive graphene strain sensor for sound signal acquisition and recognition. Nano Research. 2015;**8**:1627

Chapter 6

Using Wearable Devices in Educational Assessment: Smartphone Exams

Oytun Sözüdoğru and Nazime Tuncay

Abstract

We are residing in a planet where technology is contemporary in our life routines. Today, smartphones are one of the vastest revolutions in individuals' lifespans. Smartphones are becoming increasingly popular, both in formal and informal educational environments. This chapter discusses the benefits and obstacles in using smartphones as an assessment tools and it compares the achievement of exams delivered via smart phones to paper-based exams. The result of the study indicates that; there was a significant difference between three groups of English Paper Exams, however there was not any significant difference between these groups on English Language Mobile Exams.

Keywords: mobile exam, students, English exam, assessments, success

1. Wearable technology

Wearable technology is a group of devices that can be worn by people and track and communicate the colorful information with the outside world. The first known wearable computing device was invented in 1961 by MIT Edward Thorp and Claude Shannon, and the world's first calculator wristwatch was released in 1975. Wearable technology specially Fitbit, smart watches, and smart phones is attracting more interest of many consumers in the finance, gaming, health, music fields, as well as educators specially after the 2010s. They are specially designed to address the majority of the population who are still inactive [1]. These devices can be integrated into clothing, recognizable personal accessories (glasses, contact lenses, and watches), or additional devices (pocket device to count steps) [2]. Revenues in this segment are forecast to grow even faster than unit shipments, more than tripling in value to over $32 billion by 2019 up from $10 billion in 2013 (see **Figure 1**) [3].

As wearable devices become smaller, inexpensive, and more feature packed, the opportunity for use in various applications grows alongside [4]. People have tactile and kinesthetic senses to feel the objects' properties like its size, shape, weight (light or heavy), and temperature (hot or cold), and these ensures them about the existence and the reality. In this meaning, using wearable devices in education motivates students more than the other devices.

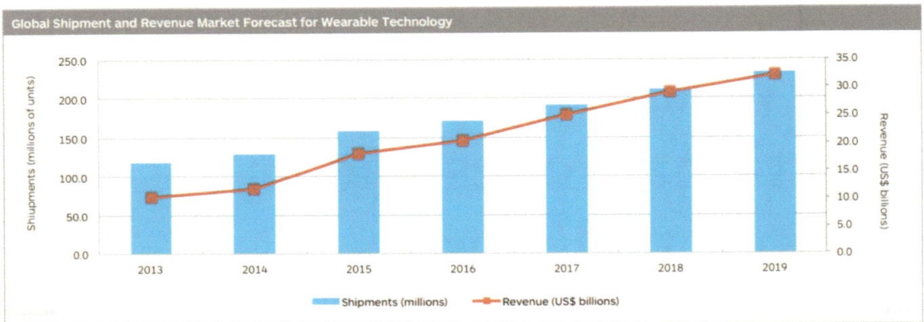

Figure 1.
Global shipment and revenue market forecast for wearable technology [3].

2. Switch to wearable devices in education: smartphone case

We are residing in a planet where technology is contemporary in our life routines. The achievement of personal goals of needs leads the individual to attain the activities voluntarily [5–8] which is also necessary to achieve their goals [5, 9]. These lead to the result that students and their motivation are the most significant part of the achievement of our courses. In this sense, choosing the best technological device brings the best possible outcome!

Today, smartphones are one of the vastest revolutions in individuals' life spans. They give mobility and excitement to its users that these modern technological devices become the most significant part of many people's lives. From online banking to watch the news on TV, we are confronting the progressions and affects that convey to our lives. The school could not stay out of these progressions, and a range of classrooms had been altered, from special spaces for the perusing of scholarly messages, to sight and sound spaces, where the utilization of data and correspondence innovation had accomplished incredible significance. Students of the twenty-first century prefer the lightest, the simplest, and the most popular way of communal and educational communication. They record everything in their smartphones for future use and are not volunteers for paper works.

Lots of students at universities have smartphones and are using its facilities like taking pictures, recording videos, and using social media. According to eminent pedagogy expert Scott P. Simkins, as far as technological innovations are alarmed, it is not pedagogy itself that mattered, but how pedagogic innovation is exploited by taking into account the specific environment in which it is implemented [10]. In the educational model where education process is carried out fully or partially with mobile technologies, students use mobile devices in wireless environments and engage in formal and informal learning [11–13]. Mobile learning model is also differentiated from other learning models by its mobility [14]. Universities and institutions have been utilizing advances, for example, synchronous videoconferencing (SV), online courses, and other kinds of technological innovations to convey language courses for the part of their educational modules. This is an open door, which is constantly important to the quickly developing requirement for understudies to end up able in utilizing innovative applications and comprehend the part of learner-focused engagement in language learning [15]. With backing of such innovations, most of these language related courses have begun being delivered online. The use of mobile phone is very popular these days specially in language learning. Mobile devices helps language teachers to use a variety of teaching methods and techniques according to students' different needs, interests, motivations and learning styles. While there have been many researches on using or integrating the mobile technology into language teaching in literature, very few of them is about the devices in wireless environments [11, 12]. The most important difference between mobile learning and other learning activities is that learners are continually on the move [14]. Universities and institutions have been utilizing advances, for example, synchronous videoconferencing (SV), online courses, and other kinds of technological innovations to convey language courses for the part of their educational modules. This is an open door, which is constantly important to the quickly developing requirement for understudies to end up able in utilizing innovative applications and comprehend the part of learner-focused engagement in language learning [15]. With backing of such innovations, most of these language-related courses have begun being delivered online. Nowadays, the use of mobile phone has received considerable attention in education as well as in language learning. Language teachers use a variety of teaching methods and techniques by considering students' different needs, interests, motivations, learning styles, and strategies as well as their pace in learning. While there have been many researches on using or integrating the mobile technology into English language teaching in literature, very few of them dwell on the usefulness of smartphones as an assessment tool from students' perspectives. Several studies have investigated the impact of mobile phones on learning outcomes in adult learning programs among rural populations and poor communities in developing countries [16, 17]; and some examined the use of mobile devices to support intentional informal learning among experienced users [18]. Ranieri and Bruni [19] stated that mobile phones are used for storytelling as well. Ranieri and Pachler [20] delivered a research study and collected data through formal and informal meetings, direct and indirect observations, interactions with participants, and focus groups and concluded in their research study that adults have great trust in the power of the media but were somewhat disappointed at their own lack of skills.

Mobile education has been delivered to university students for decades, and lots of researchers have delivered researches to discuss its efficiency and students' perspectives about it [21–25]. However, using mobile technologies like smartphones in education is relatively a new concept, and several educators and researchers start discussing this new technology in their reports [26–30]. There are lots of portable equipments like smartphones m-learning feasible at anytime and anyplace compared to the use of a notebook that can easily be damaged and does not last long [26–28]. Some research studies among Islamic education teachers are delivered for using mobile phones in secondary schools, and it is found that there is a potential for m-learning produced for Islamic education in secondary schools [29, 30].

Mobile phones have been used to provide access to contextually relevant information in clinical education [31], to create digital narratives to be used in adult education [32], and as vehicles for interactive museum guidebooks [33]. There are also studies which have been focused on developing assistive, mobile, experiential language learning applications to enhance daily literacy education anywhere and at anytime [34, 35]. Some researchers stressed that mobile media are commonly exploited in both more and less conscious modes [36–38]. Jankovića [39] examined the simultaneous impact of Facebook and smartphone usage on leisure activities and college adjustment of students in Serbia. Rheea and Kimb [40] delivered a survey with a total of 450 workers in Korea to see if there were differences in the effects of breaks with smartphones (e.g., browsing the Internet or using social network services) which have a different association with "conventional breaks" (e.g., walking or chatting face-to-face with friends).

3. Some problems on the way

There are some obstacles in using mobile phones in education. Some of these are students not having smartphone, slow Internet access, and insufficient smartphone usage awareness. Also, administrators', teachers', and students' negative attitude toward smartphone usage in education may present an obstacle in this new technology's usage. Some research studies show that there was no significant difference between the traditional, blended, and mobile groups of students' paper and mobile exam results [41]. Also, studies show that there is not a significant difference between male and female students' exam results [41]. The only difference was between the students who were familiar with the mobile technology exams and between those who were novice.

Students' perspectives are vital guides for upcoming directions in teaching and learning [22]; therefore, research studies aimed first by finding students' attitudes to smartphone usage; delivered education to three different groups of students to measure if there is any significant difference between students having mobile courses or other courses; and then the research is directed to students' perspectives of mobile education. The increasingly widespread use of new communication methods via smartphones occupies an important place in the lives of young people and influences their leisure activities [39, 42]. A high percentage of students at universities have the latest technology smartphones and are professionals using its facilities like taking pictures, creating albums, and using Gmail, Viber, WhatsApp, and Facebook perfectly with their phones. Due to the reasonable price of mobile Internet connection plans, this usage increases day by day. Smartphones are today's handheld computers for configuring the daily schedules, saving large documents, watching videos, listening music, using Internet, using World Wide Web, video conferencing, and much more than a human mind can imagine.

There are also some researchers which state that using smartphones in a classroom is supported by data suggesting that the use of such technology (e.g., text messaging during class) is negatively related to sustained attention, and sustained attention itself is positively related to academic performance. On the other hand, the use of mobile technologies in the classroom also stimulates students while they learn new material [43]. Due to strong mobile technology infrastructure in communication and Internet connections, students can benefit from both formal and informal learning methods [44]. Sometimes, social media tools may not fit into the configurations of all mobile devices. Some of the functions may be disabled, and frequent update of software is required [45].

4. Using wearable devices in assessment: mobile assessment

Computer-assisted learning environments made use of branching based on learner interactions that were the same for all learners in that same situation [46], and mobile phone-assisted learning environments take learning a step forward. Mobile phones are being used in a variety of assessing purposes. Self-assessment and peer assessment can be meaningful forms of formative feedback [47]. It is critical to a teacher's ability to adapt lessons and check for student understanding [48]. Using suitable technology for a successful implementation is important for assessing students' performance about the key concepts related to the unit [49, 50], and in a current research, a new method of assessment via smartphones is used. Smartphone exams are being used in some universities for assessments. **Figure 2** is a screenshot of two pages in an exam (the first page and the last page).

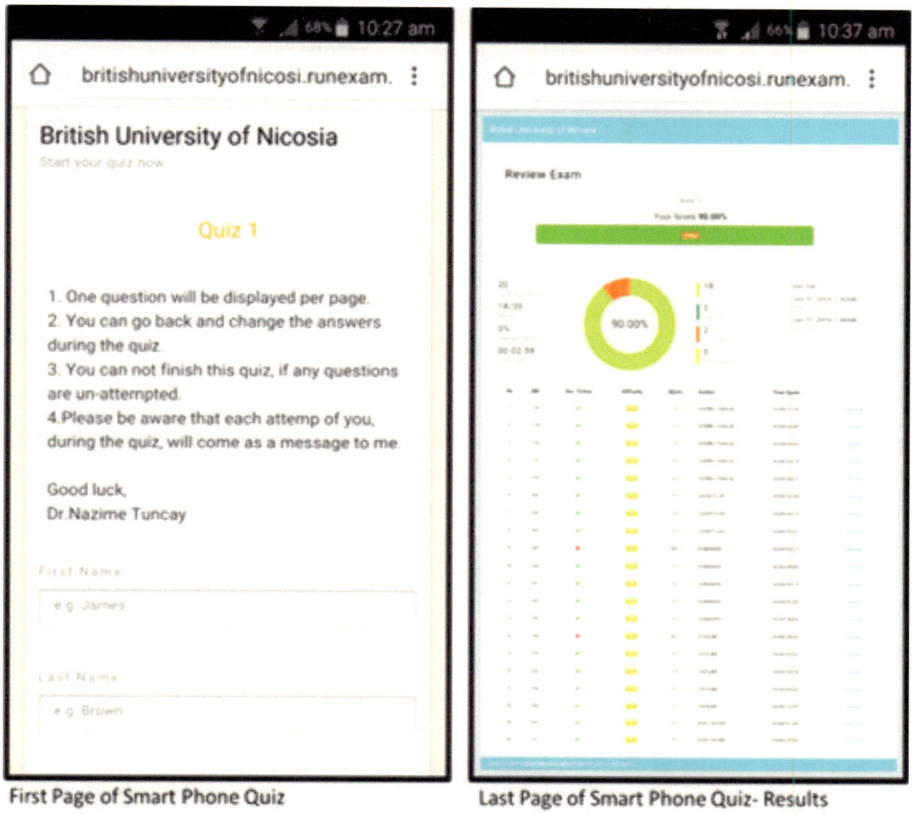

Figure 2.
An exam with smartphones [41].

In the exam, students and teachers were getting their results as soon as they finish the exam. This method claims to be the fastest method of assessing and evaluating students' progress.

5. A recent research with smartphones

Three groups of students have attained to the smartphone exams at Cyprus Science University. The first group of students are the students who were liking coming to classroom and listening to teachers in the classroom. The second group of students were preferring to come to some of the courses and to follow the other courses from mobile technologies. The third group of students were mostly working and were not able to come to class; therefore, they were following the courses from their smartphones. This research is based on a qualitative research design that meanings, perceptions, and awareness of the prospective teachers have a potential impact to retrieve the qualitative findings within an inductive process.

Seventy-five volunteer students who enrolled English I course in Guidance and Psychological Counseling program became part of this research. Volunteer participation provided a ground for confidentiality and trustworthiness within the process. In this research, trustworthy mobile phones and mobile exam programs were used as instrument tools. The mobile exam questions were distributed to students on the exam time. Students who took the exam and teachers who were the invigilators during smartphone exams had been given special training about how to use smartphones and how to access to the mobile exam via smartphones. Therefore, there were not any problems with the usage of smartphones.

At the end of mobile exams, the average scores of mobile and paper exams were compared, and students' results were driven from these comparisons. Blended course students' English paper exam results (M = 87.76; SD = 12.81) were higher than the English mobile course students' English paper exam results (M = 84.48; SD = 14.44), which was higher than the traditional course students' paper exam results (M = 83.24; SD = 14.60). Traditional course students' English mobile exam results (M = 73; SD = 16.46) is higher than the blended course students' English mobile exam results (M = 72.60; SD = 23.14), which is slightly higher than the mobile course students' English mobile exam results (M = 72.53; SD = 19.28). These results can be seen in **Table 1**.

Students had *English paper exams* in three different classes: traditional, blended, and mobile. There was a normal distribution between the marks and an equal number in three groups; one-way ANOVA was used to check if there was a meaningful difference between these three groups. According to the results of this test, there was not a significant difference between the three groups of $F(2, 72) = 1.86$, $p = 0.16$. The achievement of students in traditional, blended, and mobile classes in *English mobile exams* was also calculated statistically. There was not meaningful significant difference between three groups on $F(2, 72) = 0.53$, $p = 0.95$ (see **Table 2**).

These results can be interpreted as mobile exams which are also possible in education, and it's just a choice of the examiners whether they want to make paper exams or mobile exams.

Paper exam results of the two courses are used to make comparisons. In these comparisons, students belonging to three different groups are taken into consideration.

Blended course students' *English paper exam results* (M = 87.76; SD = 12.84), which are higher than the mobile course students' *English paper exam results* (M = 84.48; SD = 14.44), are higher than the *English paper exam* results of traditional class (M = 83.24; SD = 14.60).

Traditional course students' *computer paper exam* results (M = 94.44; SD = 5.88) are higher than the blended course students' *computer paper exam* results (M = 89.04; SD = 11.17), which are slightly higher than the mobile course students' *computer paper exam results* (M = 86.76; SD = 14.23). These results can be seen in **Table 3**.

Blended course students' English paper exam results and *traditional course students' computer paper exam results* were the highest among the students' groups.

		N	Mean	Std. deviation	Std. error
English paper exams	Traditional	25	83.24	14.60	2.92
	Blended	25	87.76	12.81	2.57
	Mobile	25	84.48	14.44	2.89
	Total	75	85.16	13.96	1.60
English mobile exams	Traditional	25	73.00	16.46	3.29
	Blended	25	72.60	23.14	4.63
	Mobile	25	72.00	18.43	3.69
	Total	75	72.53	19.28	2.27

Table 1.
English paper exams and English mobile exams.

		Sum of squares	Df	Mean square	F	Sig.
English paper exams	Between groups	706.16	2	353.080	1863	0.163
	Within groups	13643.92	72	189.499		
	Total	14350.08	74			
English mobile exams	Between groups	40.67	2	20.33	0.053	0.948
	Within groups	27478.00	72	381.64		
	Total	27518.67	74			

Table 2.
One-way ANOVA results for English paper exams and English mobile exams.

		N	Mean	Std. deviation	Std. error
English paper exams	Traditional	25	83.24	14.60	2.92
	Blended	25	87.76	12.84	2.57
	Mobile	25	84.48	14.44	2.89
	Total	75	85.16	13.93	1.61
Computer paper exams	Traditional	25	94.44	5.88	1.18
	Blended	25	89.04	11.17	2.23
	Mobile	25	86.76	14.23	2.85
	Total	75	90.08	11.31	1.30

Table 3.
English and computer paper exams.

A one-way ANOVA was conducted to compare the *English paper exam results and computer paper exam results of traditional, blended, and mobile groups of students.* There was a significant effect of three groups of $F(2, 72) = 0.70$, $p = 0.50$ in English paper exams as well as computer paper exams.

$F(2, 72) = 3.23$, $p = 0.046$ (see **Table 4**).

Although there was a meaningful difference in paper exams of computer and English courses; there was not any significant difference in their mobile exams as it can be seen in **Table 5**.

Three groups of students have attained to mobile, blended, and traditional courses for 3 months. Results of the questionnaires bring out the conclusion of compatibility and standardization. The results show that students are as good at paper exams as they are at mobile exams. Therefore, we can conclude that smartphones can be used as assessment tools in mobile English exams and the choice does not affect the students' success at the end-of-course exams. This gives a huge flexibility to the courses and freedom to teachers and students. The positive side of using smartphones is for teachers, who do not need to grade numbers of exam papers at the end of each exam. Neither should they have huge amounts of papers for examinations; thus, they save time and money. When we integrate mobile learning environments into our classrooms, teachers are required to know how to use and support that technology [51]. This may be a negative side in a smartphone usage for some teachers. Some of the limitations of this study are that it assumes that there is not an effect of sex on the results and it is restricted only with 75 first form psychology department students. Further studies about this can also be delivered to measure effect of mobile exams on other courses and with different groups of students.

		Sum of squares	Df	Mean square	F	Sig.
English paper exam results	Between groups	272.72	2	136.36	0.70	0.50
	Within groups	14077.36	72	195.52		
	Total	14350.08	74			
Computer paper exam results	Between groups	777.84	2	388.92	3.23	0.046
	Within groups	8681.68	72	120.58		
	Total	9459.52	74			

Table 4.
One-way ANOVA results.

		Sum of squares	Df	Mean square	F	Sig.
Computer mobile exams	Between groups	60.67	2	30.33	0.24	0.784
	Within groups	8939.00	72	124.15		
	Total	8999.67	74			
English mobile exams	Between groups	40.67	2	20.33	0.053	0.948
	Within groups	27478.00	72	381.64		
	Total	27518.67	74			

Table 5.
One-way ANOVA results for computer mobile and English mobile exams.

This research was significant in its own ways of research and its findings; and it aims to compare the success of students in English paper exams and English mobile exams, as well as discussing smartphone pros and cons as assessment tools.

5.1 Internet access problems

From the previous experiences, it was observed that there were Internet accessibility problems when all the students tried to access the exam at the same time. Therefore, students are divided into groups and entered the exam. Even with smaller groups, it was observed that the questions were emerging slowly. By increasing the speed of Internet access, this problem was elevated.

5.2 Print screen and copy problems

Mostly, students had tried to find a way to cheat or to disobey the given rules; and they tested the programs by their own ways. They tried to take screenshots of the program, and this was prevented successfully. The students who tried to do this were warned by the course teacher. One student tried to shade the questions and cheat; this was also successfully prevented by displaying him a warning message.

5.3 Translation problems

The exam started when the teacher had made an active link on the Internet. Since the students used translation programs in their daily lives, their smartphones instantly converted the exam to the students' native language, which was a critical problem for a language exam. Technicians corrected the problem quickly; however, it was a nasty thing. There are several possible solutions to this problem: additional codes can be included to the exam software to prevent such a case; the software can be distributed to students offline and can be programmed to send the results to the teachers' smartphone. Another possible solution to this may be instead of distributing exam papers to the students, teachers can distribute smartphones with restricted facilities to students, and they can collect these at the end of the exams to be used for future exams.

6. Wearable devices in future

We are residing in a planet where technology is contemporary in our life routines. The more that you know, the more that you want to know! Knowledgeable people are generally more keen on learning new technological devices. People's relatively high rates of prior experience with computers and smartphones may partially explain the sample's high willingness to accept smart wearable devices [4]. Today, smartphones are one of the vastest revolutions in individuals' life spans. Smartphones are becoming increasingly popular, both in formal and informal educational environments. Although benefits and obstacles in using smartphones as assessment tools can be discussed, "70 percent of students and teachers agree that they prefer to write work and notes on their computers rather than writing on paper" [52], and recent studies shows that students are as successful in smartphone exams as they are in written exams.

There are different students with different social needs: some are keen on being virtually social, and some are keen on being physically social (see **Figure 3**). Some research studies show that the younger physically social students are more successful than the younger virtually social ones [53, 54]; a solution to these would be improving wearable technologies in a way that students can both be physically and virtually social!

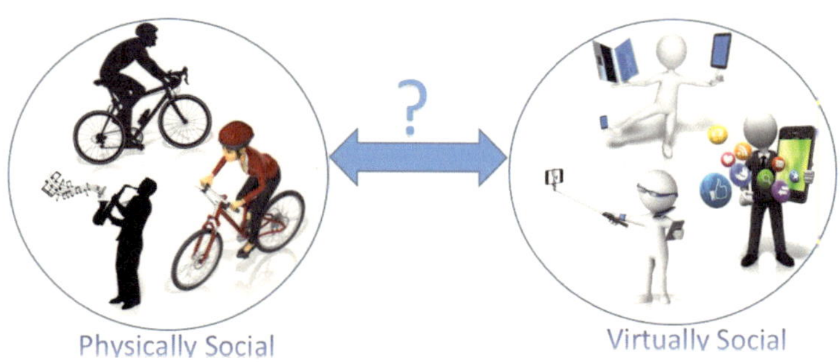

Figure 3.
Physically virtual and virtually social [53, 54].

Author details

Oytun Sözüdoğru[1*] and Nazime Tuncay[2]

1 University of City Island, Famagusta, Northern Cyprus

2 Cyprus Science University, Kyrenia, Northern Cyprus

*Address all correspondence to: oytun.sozudogru@adakent.edu.tr

References

[1] Noah A, Spierer DK, Gu J, Bronner S. Comparison of steps and energy expenditure assessment in adults of Fitbit Tracker and Ultra to the Actical and indirect calorimetry. Journal of Medical Engineering & Technology. 2013;**7**(7):456-462. Available from: https://www.ncbi.nlm.nih.gov/pubmed/24007317

[2] Shantz JAS, Veillette C. The application of wearable technology in surgery: Ensuring the positive impact of the wearable revolution on surgical patients. Frontiers in Surgery. 2014;**1**(39):1-4. Available from: https://www.ncbi.nlm.nih.gov/pmc/articles/PMC4286964/pdf/fsurg-01-00039.pdf

[3] Walker S, Roashan R. Wearable Technology: The Small Revolution is Making Big Waves. 2018. Available from: https://technology.ihs.com/api/binary/526640 [Accessed: December 15, 2018]

[4] Puri A. Acceptance and usage of smart wearable devices in Canadian older adults, a thesis presented to the Public Health and Health Systems. Waterloo, Ontario, Canada: University of Waterloo; 2017

[5] Goulimaris D. The relation between distance education Students' motivation and satisfaction. Turkish Online Journal of Distance Education. 2015;**16**(2)

[6] Harrison JM, Blakemore CL, Buck MM, Pellett TL. Instructional Strategies for Secondary School Physical Education. 4th ed. Dubuque, IA: Brown & Benchmark; 1996

[7] Hoy WK, Miskel CG. Educational Administration: Theory, Research, and Practice. 2nd ed. New York: Random House; 1982

[8] Weinberg RS, Gould D. Foundations of Sport and Exercise Psychology. 3rd ed. Champaign, IL: Human Kinetics; 2003

[9] Robbins SP. Organisational Behaviour: Concepts, Controversies and Applications. NJ: Prentice Hall; 1998

[10] George SB. Towards Innovative Pedagogy. The Hindu: Education Plus; 2014. Available from: http://www.thehindu.in

[11] Litchfield AJ, Dyson LE, Lawrence E, Zmijewska A. Directions for m-learning research to enhance active learning. In: ICT: Providing Choices for Learners and Learning: Proceedings Ascilite Singapore 2007. 2007. pp. 587-596

[12] Oran KM, Karadeniz Ş. Internet tabanlı uzaktan eğitimde mobil öğrenmenin rolü. In: Academic Bilişim'07—IX. Academic Computing Conference Proceedings; 31 January-2 February 2007. Kütahya: Dumlupınar University; 2007

[13] Vavoula G, Sharples M. Lifelong learning organisers: Requirements for tools for supporting episodic and semantic learning. Journal of Educational Technology & Society. 2009;**12**(3):82-97

[14] Sharples M, Taylor J, Vavoula G. Towards a theory of mobile learning. In: Proceedings of m Learn 2005 Conference; Cape Town, South Africa. 2005

[15] Hoven D. Designing for disruption: Remodelling a blended course in technology in (language) teacher education. In: Proceedings of the 23rd Annual Conference of the Australasian Society for Computers in Learning in Tertiary Education "Who's Learning? Whose Technology?". Sydney: Sydney University Press; 2006. pp. 339-349

[16] Aker JC, Ksoll C, Lybbert TJ. Can mobile phones improve learning? Evidence from a field experiment in

Niger. American Economic Journal: Applied Economics. 2012;**4**(4):94-120

[17] Balasubramanian K, Thamizoli P, Umar A, Kanwar A. Using mobile phones to promote lifelong learning among rural women in southern India. Distance Education. 2010;**31**(2):193-209

[18] Clough G, Jones AC, McAndrew P, Scanlon E. Informal learning with PDAs and smartphones. Journal of Computer Assisted Learning. 2008;**24**(5):359-371

[19] Ranieri M, Bruni I. Mobile storytelling and informal education in a suburban area: A qualitative study on the potential of digital narratives for young second generation immigrants. Learning, Media and Technology. 2013;**38**(2):217-235

[20] Ranieri M, Pachler N. Inventing and re-inventing identity: Exploring the potential of mobile learning in adult education. Prospects. 2014;**44**(1):61-79

[21] Bell B. Theorising Teaching in Secondary Classrooms: Understanding our Practice from a Sociocultural Perspective. New York, NY: Routledge; 2011

[22] Forbes D, Khoo E. Voice over distance: A case of podcasting for learning in online teacher education. Distance Education. 2015;**36**(3):335-350. DOI: 10.1080/01587919.2015.1084074

[23] Khaddage F, Müller W, Flintoff K. Advancing mobile learning in formal and informal settings via mobile app technology: Where to from here, and how? Educational Technology & Society. 2016;**19**(3):16-26

[24] Salmon G, Mobbs R, Edirisingha P, Dennett C. Podcasting technology. In: Salmon G, Edirisingha P, editors. Podcasting for Learning in Universities. New York, NY: Open University Press; 2008. pp. 20-32

[25] Garrison DR, Anderson T, Archer W. Critical inquiry in a textbased environment: Computer conferencing in higher education. The Internet and Higher Education. 2000;**2**(2-3):87-105

[26] Nawi MAM, Jamsari EA, Hamzah MI, Sulaiman A, Umar A. The impact of globalization on current Islamic education. Australian Journal of Basic and Applied Sciences. 2012;**6**(8):74-78

[27] Shuib AS. Reka Bentuk Kurikulum M-Pembelajaran Sekolah Menengah: Teknik Delphi. In: Proceedings of Regional Conference on Knowledge Integration in ICT. 2010. pp. 652-665

[28] Wagner ED. Enabling Mobile learning. Educause Review. 2005;**40**(3):40-53

[29] Nawi A, Hamzah MI, Rahim AAA. Teachers acceptance of Mobile learning for teaching and learning in Islamic education: A preliminary study. Turkish Online Journal of Distance Education. 2015;**16**(1):184-192

[30] Nawi MAM, Hamzah MI. Mobile Fatwa (M-Fatwa): The Integration of Islamic fatwa through mobile technology. Turkish Online Journal of Distance Education. 2014;**15**(2):108-116

[31] Hann S. About family violence networks in New Zealand. 2012. Available from: https://nzfvc.org.nz/sites/nzfvc.org.nz/files/ABOUT%20FAMILY%20VIOLENCE%20NETWORKS%20IN%20NZ.pdf

[32] Herrington A. Using a smartphone to create digital teaching episodes as resources in adult education. In: Herrington J, Herrington A, Mantei J, Olney I, Ferry B, editors. New Technologies, New Pedagogies: Mobile Learning in Higher Education. Wollongong: University of Wollongong; 2009. pp. 28-35. Available from: http://ro.uow.edu.au

[33] Sung YT, Chang KE, Hou HT, Chen PF. Designing an electronic guidebook for learning engagement in a museum of history. Computers in Human Behavior. 2010;**26**:74-83

[34] Lumsden J, Leung R, D'Amours D, McDonald D. ALEX_: A mobile adult literacy experiential learning application. International Journal of Mobile Learning and Organisation. 2010;**4**(2):172-191

[35] Munteanu C, Molyn H, McDonald D, Lumsden J, Leung R, et al. "Showing off" your mobile device: Adult literacy learning in the classroom and beyond. In: Proceedings of the 13th International Conference on Human Computer Interaction with Mobile Devices and Services. New York: ACM; 2011. pp. 95-104

[36] Bayer JB, Campbell SW. Texting while driving on automatic: Considering the frequency-independent side of habit. Computers in Human Behavior. 2012;**28**(6):2083-2090

[37] Farman J. Mobile Interface Theory. New York, NY: Routledge; 2012

[38] Humphreys L, Von Pape T, Karnowski V. Evolving mobile media: Uses and conceptualizations of the mobile internet. Journal of Computer-Mediated Communication. 2013;**18**(4):491-507

[39] Jankovića B, Nikolićb M, Vukonjanskib J, Terekb E. The impact of Facebook and smart phone usage on the leisure activities and college adjustment of students in Serbia. Computers in Human Behavior. 2016;**55**:354-363

[40] Rheea H, Kimb S. Effects of breaks on regaining vitality at work: An empirical comparison of 'conventional' and 'smart phone' breaks. Computers in Human Behaviour. 2016;**57**:160-167

[41] Tuncay N. SmartPhones as tools for distance education. Journal of Educational and Instructional Studies in the World. 2016;**6**(2):3

[42] Kuo T, Tang HL. Relationship among personality traits, Facebook usages, and leisure activities—A case of Taiwanese college students. Computers in Human Behavior. 2014;**31**:13-19

[43] Pachler N, Cook J, Bachmair B. Appropriation of mobile cultural resources for learning. International Journal of Mobile and Blended Learning. 2010;**2**(1):1-21

[44] Rau PP, Gao Q, Wu L. Using mobile communication technology in high school education: Motivation, pressure, and learning performance. Computers & Education. 2008;**50**(1):1-22

[45] Au M, Lam J, Chan R. Social media education: Barriers and critical issues. In: Technology in Education. Transforming Educational Practices with Technology. Berlin, Heidelberg: Springer; 2015. pp. 199-205

[46] Wauters K, Desmet P, Van den Noorgate W. Adaptive item-based learning environments based on the item response theory: Possibilities and challenges. Journal of Computer Assisted Learning. 2010;**26**:549-562

[47] Spector JM, Ifenthaler D, Samspon D, Yang L, Mukama E, Warusavitarana A, et al. Technology enhanced formative assessment for 21st century learning. Educational Technology & Society. 2016;**19**(3):58-71

[48] Davis MR. Technology Fed Growth in Formative Assessment. Education Week. 2015. p. 11. Available from: http://www.edweek.org/ew/articles/2015/11/11/technology-fed-growth-in-formative-assessment.html

[49] Gregory J, Lokuge Dona K. MOOCs for Online Professional Development. Newsletter of the Business/Higher Educational Round Table, Issue 33. 2015. pp. 9-10. Available from: http://www.bhert.com/publications/newsletter/BHERTNews33.pdf

[50] Gregory J, Lokuge Dona K, Baily M. Blended learning—Flexibility, increased interaction to keep students engaged. In: Paper Presented at the Innovate and Educate: Teaching and Learning Conference 2015. Adelaide: Australia; 2015

[51] Kukulska-Hulme A, Shield L. An overview of mobile assisted language learning: From content delivery to supported collaboration and interaction. ReCALL. 2008;**20**(3):271-289

[52] Tuncay N. The influence of the "approach gap" between students' and teachers' e-learning preferences. In: Proceedings of the International Conference on E-Learning. 2013. pp. 488-494. Available from: http://proxy-remote.galib.uga.edu/login?url=http://search ebscohost.com/login.aspx?direct=true&db=ehh&AN=91956471&site=eds-live

[53] Tuncay N. MLMDS: Traditional and Online Mathematics Courses with Dyslexic Students. Ankara: Çözüm Educational Publishing; 2018

[54] Tuncay N. Social Media Ban: Virtually Social or Physically Social? Ankara: Çözüm Educational Publishing; 2018

Chapter 7

Wearable Devices and their Implementation in Various Domains

Menachem Domb

Abstract

Wearable technologies are networked devices that collect data, track activities and customize experiences to users' needs and desires. They are equipped, with microchips sensors and wireless communications. All are mounted into consumer electronics, accessories and clothes. They use sensors to measure temperature, humidity, motion, heartbeat and more. Wearables are embedded in various domains, such as healthcare, sports, agriculture and navigation systems. Each wearable device is equipped with sensors, network ports, data processor, camera and more. To allow monitoring and synchronizing multiple parameters, typical wearables have multi-sensor capabilities and are configurable for the application purpose. For the wearer's convenience, wearables are lightweight, modest shape and multifunctional. Wearables perform the following tasks: sense, analyze, store, transmit and apply. The processing may occur on the wearer or at a remote location. For example, if dangerous gases are detected, the data are processed, and an alert is issued. It may be transmitted to a remote location for testing and the results can be communicated in real-time to the user. Each scenario requires personalized mobile information processing, which transforms the sensory data to information and then to knowledge that will be of value to the individual responding to the situation.

Keywords: wearable devices, device architecture, healthcare, visually impair, automatic navigation

1. Introduction

Wearable devices have embedded sensors which acquire the data for which they were built. This data are then pushed to its integrated processor. The processor analyzes this data and accordingly launches commands, actuators or activates other sensors to collect more data or execute tasks according to predefined scenarios and processes. To promote device standardization and quick adaptation to a wide variety of goals and purposes, we propose a three-layer architecture: the common layer, the domain layer and the special-purpose layer. The basic layer contains common elements required for any wearable device: motherboard, power supply, processor, operating system, communication ports, a grid of sockets and adapters for sensors plugin and software applications.

With the emergence of growth in various technologies, it is predicted that soon about 50 billion new devices will be added world-wide. This raises two major issues: a huge amount of data and heterogeneous devices with severe integration issues. These concerns remain when referring to wearable technology. Typical wearable body sensor networks consist of tiny, smart, low-power and self-organized sensors to observe physiological signals of a human body. Standardization, compliance, effective coexistence and interoperability among multiple technologies are required to ensure end-to-end network routing and connectivity among wearables and external devices. M. Alam et al. [4] review multi-standard and multiple technologies based wearable wireless for inter-device communication. Coexistence and inter-operability are challenges discussed along with utilization of possible technologies for on-body, body-to-body and off-body communications. It explores several schemes to ensure effective coexist among multiple technologies and issues related to interoperability.

In this chapter, we describe the architecture and its operation in several domains, one implementation per domain. Lauren Kolodzey et al. [1] reviewed 614 articles aiming to provide an objective overview of the literature about the use of wearable technology in clinical and simulated surgery. They found that applications of wearable technology mainly focused on improving the safety and efficiency of intraoperative processes. The associated applications were wide-ranging and designed for use by a variety of care providers, thereby reflecting the interconnected relationship between intraoperative safety and the entire healthcare team. It suggests that wearable devices resolve certain human factors that negatively influence performance and safety in the operating room. For example, a display of patient variables to mitigate conflicts associated with patient care tasks and the distracting operative environment. It recommended the use of a variety of wearable devices, such as special glass for its lightweight construction, user-friendly interface and potentially for hands-free control, special camera for capturing precise anatomical details.

The rest of this chapter is composed as follows. In Section 2, we describe the platform and technology components used for developing and implementing wearable-based systems. In Section 3, we outline wearables in the healthcare domain, which is the most advanced domain and with the highest number of production implementations. In Section 4, we review several wearable implementations in several domains, such as agriculture, cconstruction and others. In Section 5, we describe in detail our original implementation of a wearable-based system assisting visually impaired people in safety walking through and avoiding obstacles. At Section 6, we conclude and outline potential directions for further advancements in this subject.

2. Technology enablers

S. Park et al. [2, 3] explore advanced wearables and accordingly recommend guidelines for successful development and deployment of comprehensive wearable systems. Among these are the use of a variety of sensors, each sensor should be flexible, adaptive, effective and reliable.

2.1 Variety of sensor types and flexible, effective and practical sensors

Sensors are needed for capturing various aspects and parameters to be handled simultaneously [4], such as vital signs sensors working at the same time and having multiple sensor types: heart rate, body temperature, pulse oximetry, blood glucose level of different types, the number of sensors may change to capture the signals required to compute a single parameter. In some cases, it may require placing the sensors is specific locations, for example, electrocardiography for recording the electrical activity of the heart where the sensors are placed in three locations on the body. In addition, sensors should be easy for attachment and removal, or for plugging and playing, as sensors may be used at different times and changing requirements. In most cases, parallel processing is required. For example, a pilot during a flight-simulation wants to analyze his overall body reaction during the simulation action. This requires placement of several sensor types, changing locations and types during the simulation. Practically, sensors should be low cost, lightweight, adaptable to the wearer body, distributed power supply and data communication among sensors and processes in the wearable network.

The concept of packaging and fabrication technologies has been widely used and keeps improving with new various materials [5]. These developments enable embedding sensors, such as gyroscopes, accelerometers, camera, motion sensing, physiological and biochemical sensing, into a rigid and flexible platform, adding capabilities to wearable devices. Mobile devices have been integrated with wireless communication technology. The constant growth of broadband wireless networks opens a new era for wearable devices and sensors to continuously monitor the health of patients remotely.

2.2 The generic paradigm for connecting wearables

M. Alam and Ben Hamida [6] propose a generic paradigm, which can serve as a platform for many existing and future applications, such as healthcare, disaster recovery, people safety and more. The key advantage is its wearable Wireless Body Area Networks (WBANs) capabilities, enabling remote and ad hoc deployment of networks. Envisioned applications in this context, range from the popular medical field, continue with entertainment, lifestyle, gaming and ambient intelligence. Applications, such as disaster recovery, rescue, safety, wearable technology can also play a role to protect critical and valuable assets. The network is designed in such a way that the coordinating device communicates with implanted and on-body

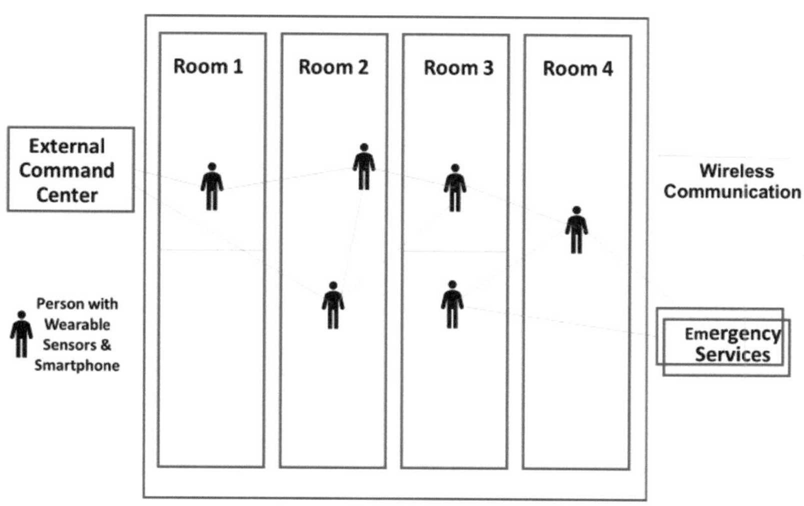

Figure 1.
The generic paradigm.

sensors, transmits the collected information to a remote monitoring station. **Figure 1** depicts the advantages of wearable communications to enhance the readiness and alertness of the wearers, devices and vehicles to act as an integrated unit, regardless of their physical location and distance from the occurrence. Enhancing this composition, we may implant cameras in the wearables and provide real-time information to all who involved in a given situation. We may assume that this architecture is the right architecture and infrastructure for any wearable-based functions.

3. Wearables for health

A dominant area of wearables is health. It is aiming to predict and treat common cases by acquiring and processing physiological and environmental data. Wearable technologies allow consumers to be better at converting personal, biological and environmental data into valuable consumer insights. Wearables can transmit the data to and from the consumer at the appropriate time, creating new consumption experiences that can improve the landscape of health and fitness. These insights may turn into holistic decisions and goal-directed actions, especially if patients allow the access to their physiological data, collected from wearables. A new generation of wearable sensors enables physicians to capture long-term-patients' activity levels and exercise compliance, facilitating effective dispensing of medications for chronic patients and provide tools to assess their ability to perform specific motor activities, and propose rehabilitation solutions.

Wearables enable remote health monitoring of patients [7, 8]. The data are sent from the wearable to the physician's office, avoiding the need for office visits. The ability to continuously track patients' health helps identifying potential problems through preventive interventions and so enhances the quality of care and save money, since the cost of prevention is most cases is less than the treatment cost. The resulting higher quality of care at lower cost would also contribute to better operating efficiencies and lower overhead costs for insurance companies, as resources can be better spent on providing care and not on measures to ensure high quality of care is being provided. This is where wearables have a critical role to play in creating and serving as the core of an ecosystem essential for facilitating the seamless transformation of data to deliver value.

Healthy lifestyle improves employee's productivity and lower absence rate [9]. Insurance companies can collect its activity and sleeping data to leverage the data for personal insurance plans and reward employees for good health score. An American insurance company issued a wearables based health program pilot, which continuously collects invasive and noninvasive data, such as vital signs. Artificial intelligence provides an added value to healthcare with a focus on diagnosis, treatment, patient monitoring and prevention.

3.1 Personalization

The doctor, with the help of a software expert can quickly create a program based on the needs of the patient. Early diagnosis: precise medical parameters allow early detection of symptoms. Remote patient monitoring: healthcare professionals can monitor patients remotely and in real-time using wearable devices. Adherence to medication: help patients to take medications on time and inform medical professionals if the patient fails to adhere to medications. Information registry: the data are stored in real time allowing an exhaustive analysis of the information. The result is a complete and precise report about the patient's medical history, which can be shared with other specialists. Optimum decisions: the doctor can analyze the data to make better clinical decisions, to enhance the patient's quality of life. Saving healthcare cost: remote healthcare using wearable devices means saving time and mobility.

Recent emergence of new materials accelerates the development of non-invasive skin-based wearable devices [10], which are expected to be compatible with human skin: flexible, stretchable and less irritating, and comply with: sensitivity to changes in body temperature, changes in the body and an adequate detection limit. Following are several examples of skin-based devices in healthcare applications: predicting a sudden attack and providing the means to cope with it; detecting genetic cancer syndromes or rapid changes in heart-beat rate; early evidence of vascular events; detecting abnormal respiration rate; monitoring body temperature and biosensing clothing. Wearable strain sensors are used for detecting and monitoring of movement-based signals, such as heart-beat rate and respiration rate. It is lightweight, reliable, flexible, stretchable and aligned with the diverse healthcare applications.

3.2 Diseases

Several researchers proposed wearable-based solutions for specific diseases [11], to assist in curing or relieving the symptoms of a list of diseases as follows:

1. Sleep apnea: interruptions or a decrease in breathing for few seconds up to a minute. The treatment types depend on the severity of the case and ranges from weight loss to surgical operations. DT is a wearable oral device for following the prescribed therapy for sleep apnea. It measures the temperature, movement and head position of patients by determining the spatial orientation of the device in the mouth.

2. Chronic obstructive pulmonary disease: a common lung disease that leads to shortness of breath. An ear wearable monitors the physical activities that allow patients to continuously evaluate their condition at home. It reduces healthcare costs for patients that can be treated at home.

3. Diabetes mellitus. A chronic disease whereby the body cannot produce enough insulin, and the control of blood glucose levels is essential for diabetic patients. A wearable artificial pancreas for monitoring glucose level. It is composed of a flexible core system as a brain and three wedges for insulin delivery, glucose sensing and glucagon delivery. Another wearable to measure blood sugar levels in diabetics is the smart contact lens that Google/Verily Life Sciences owns.

4. Cardiovascular diseases: it is related to the heart, veins, venous thrombosis, heart failure and cardiac dysrhythmia. Various wearable sensors exist for providing real-time heart rate measurements, such as the wireless blood pressure wrist monitor, which monitors blood pressure in connection with a smartphone. It was shown that the accuracy of the measurements was in good agreement with the reference clinical measurements.

5. The Vega GPS bracelet is a wearable sensor for ensuring the safety of people by monitoring their location with the use of GPS and global system for mobile communications positioning. Embrace is a wristband for monitoring physiological signals in epileptic people in real time to alert family members.

6. Mosquito-borne diseases: it causes a wide range of deadly diseases, such as malaria, chikungunya, yellow fever, the Zika virus and the Ebola virus. The kite patch is a patch-type wearable that disperses volatile compounds and is worn on a shirt to repel mosquitoes.

7. Renal failure: kidney failure and chronic kidney disease. In the treatment of renal failure, dialysis is commonly used, in which kidney function is replaced by a machine. To replace dialysis, a wearable artificial kidney has been developed.

8. Skeletal system diseases: joint disorders, osteoporosis and poor posture. Using three-dimensional gyroscopes, accelerometers and magnetometers embedded into wearable sensors, the chronic pain resulting from most skeletal diseases can be treated with transcutaneous electrical nerve stimulation and by performing therapeutic exercises. Another wearable monitor uses postural variation and warns users through vibrations when they deviate from normal posture, reminding them to return to a normal posture.

9. Sunburn prevention: the ultraviolet (UV) radiation of sunlight causing wrinkles, burns, aging and even skin cancer. Wearable UV sensors, which can be worn on the arm in the form of a bracelet, armband or wristband, are used to monitor UV exposure levels with alerts for potential skin damage and safety precautions, as well as estimating vitamin D production levels.

10. Vein finding: a wearable smart glass termed Eyes-On technology enables nurses to rapidly see the veins of patients through the skin by incorporating multispectral 3D imaging and wireless connectivity.

11. Detection of stress/depression levels: wearables are used to determine the state of mind of their users. The product is a wristband that monitors heart rate variability aiming to warn the user about a rise in personal stress levels.

3.3 Nutrition and dietetics

Real-time, effective and affordable nutrition and dietetics wearable technology and sensors are an emerging field with immense opportunities and benefits to the global nutrition challenge [12, 13]. Such revolution real time, home-, work- and hospital-based rapid, accurate and cost-effective self-detection and diagnosis of direct or indirect causes or diet deficiency or excess are much needed for generating evidence-based information and knowledge for individual and vulnerable group nutritional and dietary mitigation and lifestyle adaptation through wearable sensors and technology. These can enhance evidence-based, coherent and coordinated nutrition and dietary programs and strategies to a targeted group or illness, vital in addressing malnutrition and under-nutrition public health burden. Building wearable consumers' health and fitness prognosis, prospective digital nutrition, dietetic data and database and nutrition informatics platforms. These provide a paradigm shift in engaging participatory communication among public consumers, dietetic and nutritionist professionals in improving quality interventions, management and outcomes. Assessment and understanding of nutritional and dietary needs, and potential opportunities in functional health benefits and resource development in personalized accessibility and availability of needed resources to encourage positive behavior, diet and nutrition changes.

Diet-related deficiencies are estimated at 3.5 million deaths annually. This results in rapid urbanization and food consumption patterns that require nutrition safety. The public health nutrition wearable and implantable sensors approach provides a new perspective in human and animal nutrition and dietary. They lead to reliable and effective nutrition and health interdisciplinary approaches and tackle the ever-growing local and global nutrition challenges. Modern convenient and cost-effective wearable sensors can be used to educate, track and predict energy level and advice on interventions or activities required to improve the excess or deficiency and adaptation changes from plant-derived sources in achieving balanced choices and quantities of unique fruit and vegetable phytochemical/micronutrient needs.

The effectiveness of wearable devices and fitness trackers, and mobile application on healthy life and care delivery outcomes, such as weight loss and maintenance have been documented in developed countries. Nutritional and dietary wearable technology has a critical role in contributing to nutritional and food challenges paradigm shift in Africa. It provides real time, home-, work- and hospital-based rapid, accurate and cost-effective detection, and diagnosis of nutrition/energy or diet deficiency or excess is much needed. It supports the generation of quality information and knowledge for individual, vulnerable group to national decision-making nutrition policy and guidelines, programs and interventions towards healthier lifestyle and increasing life expectancy, more productivity and wellness. Real time is required for flexible applications of smart wearable and implantable sensors are needed in providing clues into effective fitness and feeding best practices.

3.4 Body dietary and energy balance

To estimate daily total energy expenditure (TEE) using a physical activity monitor, combined with dietary assessment of energy intake to assess the relationship between daily energy expenditure and patterns of activity with energy intake. [14] An activity monitor has been used to determine the total energy expenditure, sleep duration and physical activity. The armband was placed around the left upper triceps. Energy intake was determined by evaluating all food and drink items. TEE was correlated with BMI and body weight but inversely related to sleep duration and time lying down. Multiple linear regression analysis revealed that after taking BMI, sleep duration and time spent lying down into account, TEE was no longer correlated with energy intake. Results show the extent to which body mass, variable activity and sleep patterns may be contributing to TEE and together with reduced energy intake, energy requirements were not satisfied. Hence, wearable technology has the potential to offer real-time monitoring to provide appropriate nutrition management which is more person-centered to prevent weight loss.

4. Wearables for other domains

4.1 Construction

The known high percentage of accidents occurring in the construction industry, calls for developing safety strategies. In this section, we describe personalized construction safety-monitoring applications, incorporating wearable technology. These devices predict safety performance and management practices are identified and analyzed. Awolusi et al. [15] present a variety of solutions.

Environment sensors are silicon sensors, small and embedded communication technologies, such as Bluetooth and Wi-Fi wearables. They increase the volume and precision of environmental data, such as air quality, barometric pressure, carbon monoxide, capacitance, color, gas leaks, humidity, hydrogen sulfide, temperature, light, volatile organic compounds (VOCs) and ability to realize intelligent RFID tags. There are sensors that support a broad range of emerging high-performance applications, such as navigation, barometric air pressure, humidity and ambient air temperature sensing functions. Some of these sensors are designed for wearable technologies. Workers can be monitored while doing their normal work and at the same time having the ability to see highly localized, real-time data on things like temperature. Other wearable-sensors that can be used in wearable devices are gyroscope, light sensors, noise sensors, humidity sensors, temperature sensors, gas sensors, among others.

Wearable devices have the potential to protect workers in hazardous conditions: the use of Smart headsets for monitoring truck drivers' performance to reduce accidents; Augmented Reality headset to guide workers though complex production processes or wearable devices to predict injuries and machine downtime. According to Gartner, most companies with 500+ employees already use wearables in the workplace.

4.2 Quality of life

J. Lee et al. [16] focus on the value of sustainability in human-oriented wearables and services that seek to improve the quality of life, which involves social impact and public interest. Wearables refer to the technology and its applications with a value of sustainability having a positive impact on the improvement of quality of life, social impact and the public interest. We aim to discuss how continuously evolving wearables influence positively on human life and environment through the keyword of sustainability.

A variety of wearable devices have been launched in the market to achieve various purposes with the development of sensing technologies. One typical example is an application that constantly measures movement distance and movement conditions of users over time through motion sensors that include in wrist-wearable devices and display the measured results. Moreover, measuring the intake and consumption of calories, tracking sleep, postural correction, blood pressure, and heart rate are the most fundamental applications of the current wearables field. As such, wearable applications started by quantifying various human activities (consciously or unconsciously) numerically in daily life. Over the past few years, more wearable devices have been introduced according to their purpose with increasing performance. As a result, the demand for them to quantify individual daily lives by themselves has increased. Along with this demand, more studies of the methods to improve the quality of life by analyzing individual conditions have been conducted for application in real life, which is called the quantified self. Targets whose movements are tracked include various types of personal information, such as physical activities performed and environmental information.

4.3 Monitoring social interactions

Wrist-worn wearables enable monitoring, detecting and recording interpersonal social interaction features [17]. The wrist has embedded motion sensors, accelerometers, heart rate monitors, optical sensors, skin conductivity, skin temperature and other physiological sensors. Increased synchrony of physiological measures has been shown to lead to increased perceived empathy and positive outcome.

Leveraging data from wearables for social sensing based on interpersonal synchrony. Preliminary results show that wearable data are suitable for analyzing and quantifying social dynamics. Results indicate differences in wearable sensing data during a social interaction between two people.

4.4 Agriculture

According to Afzal et al. [18], water is a vital component in plants. They measured leaf moisture using special sensors. Results showed that variations curve of the capacitance was in the form of an exponential function, y = ae bx, where y is capacitance, x is leaf moisture content, a is the linear coefficient and b is the exponential coefficient. A new adhesive sensor, sensitive to water vapor, measures leaf surface humidity and how much water is transpired by crop plants. It exhibits different levels of conductivity depending on the humidity and provides farmers with practical information on the real-time water absorption habits of their crops. The sensor is connected to a Wi-Fi device that transmits the data to the data analyzer, which then recommends the amount of water gallons to put in which parts of the field. The sensor is used for water management to accelerate the process of breeding drought tolerant for any crop.

5. Wearables for navigation and safety systems

Automatic navigation in an unknown environment raises various challenges as many cues about orientation are difficult to perceive without the use of vision. Though assisted aids, such as global positioning system (GPS), a satellite-based radio-navigation system, which help in route finding, still it fails to fulfill safety requirements. This section proposes a framework that provides accurate guiding and information on the route traversal and the topography of the road ahead. The framework is composed of technologies, such as Lumigrids, Drone, GPS, Mobile applications andCloud storage which are used to map the road surface and generate proper navigation guidance to the end user. This is done in three stages: (1) off-line mapping of the road surface and storing this information in the cloud; (2) wearable technology used for obtaining in real-time surface information and comparing it to the data on the cloud facilitating accurate and safer navigation and (3) updating the cloud information with information collected by the pedestrian.

There are many technological navigation aids but none of them focus on pedestrian paths. Banovic et al. [19] claim that travelers require detailed information about the terrain and its challenges—size, curves, hurdles, fences, changes in elevation and proposes a three-phase safe navigation system that provides surface information of the pedestrian paths and uses this information while suggesting in real time routes to the visually impaired.

Most applications use location-sensing technology, such as GPS combined with a map to locate and guide pedestrians. Sendero [20] uses smart phone's location sensing power. Trekker Breeze [21] supports orientation using a commercial GPS receiver. In another work, [22] has combined crowd sourcing with computer vision techniques to provide additional information about traffic intersections and sidewalks or arbitrary images. Few open source [23] software systems provide similar navigation instructions on points of interest like restaurants and buildings to the user using speech or Braille output. Studies say that pedestrians are positive on using technological assisted aids to guide them for navigation [24].

5.1 The process phases

The proposed navigation system consists of the following three phases: (1) terrain mapping phase, (2) pedestrian guidance phase and (3) re-mapping of the terrain based on comparative walk-thru and terrain database. In the terrain mapping phase, an unmanned aerial vehicle is made to fly over the pedestrian path. This vehicle records the GPS coordinates of the mapped region and accurately identifies the actual terrain of the underlying pedestrian path. This data are versioned and stored in a cloud. This referential database is centrally shared for the visually impaired. The terrain mapping phase is essential to initially map all the pedestrian paths and populate the cloud with data. The pedestrian guidance phase is the phase where the stored terrain-related information on cloud is combined with the regular GPS-based route finding and in real time, it is used to guide a pedestrian in navigation. A shirt mounted device assists the visually impaired in achieving this. During the walk-thru, the mounted device with the visually impaired obtains the real-time terrain information of the path ahead and compares it to the existing information on the cloud to alert of the new challenges/hazards that may have cropped up.

5.2 The navigation system

The terrain mapping phase consists of the following components: Quadcopter—unmanned aerial object and Raspberry—a microcomputer to run required image processing algorithms and save the information to the cloud. **Figure 2** depicts the components and their interconnection used in terrain mapping phase.

Lumigrids—a LED projector projecting light in the shape of grids as presented in **Figure 3**.

Lumigrids are mounted on the quadcopter and placed facing the ground. These light grids can accurately extract the terrain information of the pedestrian path as the regular arrangement of the lights grid gets distorted based on the terrain. [24] shows how lumigrids can help cyclists to understand the terrain ahead at night and keep them safe. *Camera*—placed facing the direction of ground where the lumigrids are projected. It constantly takes the images of the patterns formed by the grids and sends it for image processing. *GPS sensor* is used to obtain the GPS location of the quadcopter drone. Raspberry Pi serves as the central computing unit for all the attached sensors. It processes the captured images of the formed light grids on the ground and obtains the required terrain information.

An interesting approach can also be used to obtain the terrain-related information by using the accelerometer data of the smart phones of other visually sound

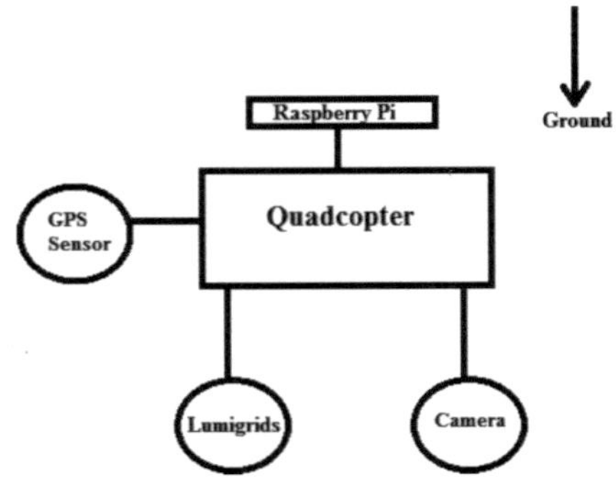

Figure 2.
Components used in terrain mapping phase.

Figure 3.
Light grids projected on ground by lumigrids projector.

pedestrians who use these pedestrian paths. The accelerometer of their mobile devices detects the vibration along the X, Y and Z-axes. The magnitude m of the acceleration is calculated as $m = \sqrt{X^2 + Y^2 + Z^2}$. This is used to predict the terrain information of the pedestrian paths.

5.3 Capturing of the terrain topography in two phases

The terrain mapping system consists of a lumigrids projector and a GPS sensor mounted on a quadcopter which flies along the pedestrian path at height "h" above ground, as depicted in **Figure 4**. The captured data are associated with its exact location [GPS], which allows the comparison between images taken from the same location. As mentioned, the process is divided into two phases. In the first phase, the terrain image and data are taken and stored in the cloud storage. To ensure accurate terrain data, while the pedestrian walks, we recapture, in the second phase, the same image from the same location.

Figure 5 describes the process of obtaining the terrain topography using lumigrids projection. The first picture on the top-left is the image of the sidewalk, we refer to in this section. The picture on the top-right presents the projection of the lumigrids projector on the sidewalk. A complete flat terrain will produce and show a perfect grid picture. However, due to some bumps in the sidewalk, as presented in the right-bottom image, some of the projected squares are distorted, representing the bump location. The resulting grid is sent to the cloud application for analysis and storing it in the cloud storage.

Figure 6 depicts the data collection and processing from the impaired person guidance perspective. It assumes the use of a smart-phone application, which continuously transmits the current person location and orientation to the cloud and obtains the data about the terrain of the path ahead. The left-top image presents the shirt with a mounted unit, which the pedestrian wears. The unit consists of a lumigrids projector, camera and a communication unit. The projector flashes on the

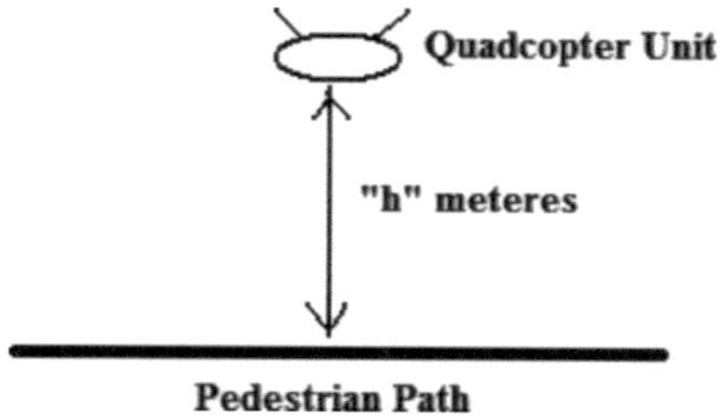

Figure 4.
The basic set-up of capturing the terrain image and its data in phase 1.

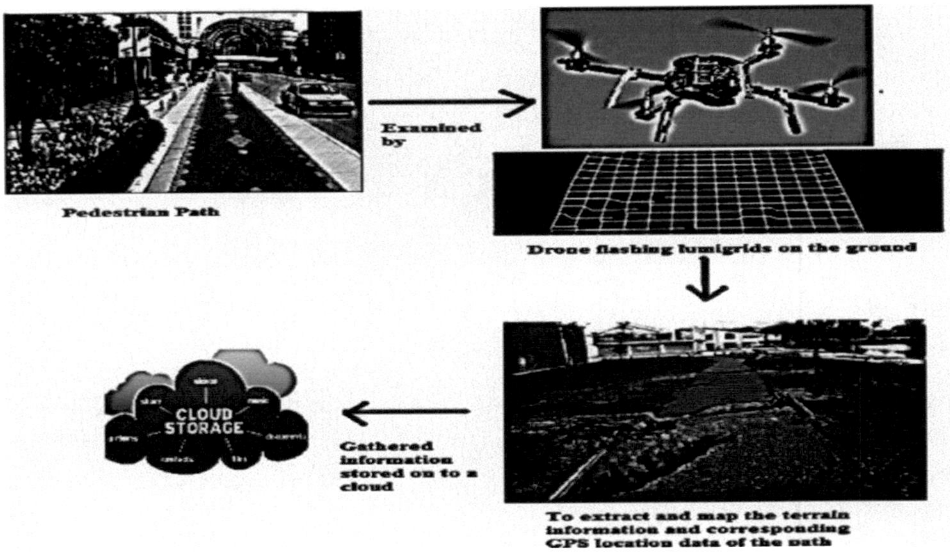

Figure 5.
The terrain mapping phase and its transmission to the cloud storage.

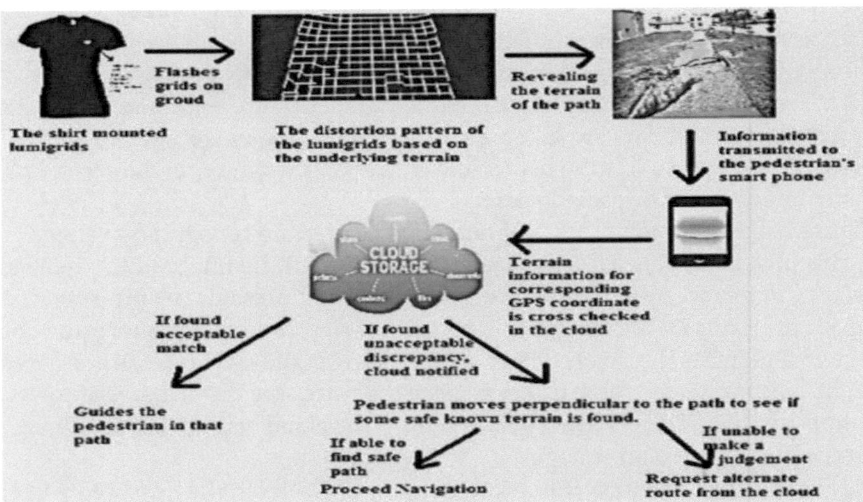

Figure 6.
The process of the pedestrian guidance phase.

ground. The camera captures the grid image formed on the ground and continuously transmits it to the smartphone application, which then transmits it to the cloud application. The application compares the received image to the already stored image and generates the most accurate image representing the terrain situation at this moment. Accordingly, the application generates the proper instructions set and sends it back to the smartphone, which guides the pedestrian accordingly. In parallel, the discrepancy between the stored data in the cloud and the data accepted from the pedestrian, is analyzed and if there is a need to update the cloud data it is done by the cloud application.

The steps of the terrain mapping phase:

1. The entire pedestrian path is divided into squares of equal area—called the sub-squares: let "k" be the area of each sub-square with side "x" which are named as (1, 1), (1, 2) and so on.

2. The height "h" is adjusted to generate the lumigrids of area "k" just enough to cover each sub square.

3. The midpoint M of the sub-square is calculated as: $M = \left(lat1 + \frac{x}{2}, long1 + \frac{x}{2}\right)$.

4. The quadcopter flying at height "h" above the ground files to the calculated M from where it flashes the lumigrids of area "k" equal to the area of the sub-square on ground. The lumigrids projector creates the light grids of dimension n × n on the ground below.

5. The following image formed on the ground shows an undistorted lumigrids of area "k" formed on an ideally flat and perpendicular surface to the quadcopter flying at a height "h" above the ground.

6. This image is captured by the mounted camera and thresholding of the input image splits the lumigrids image data from rest of the image as explained in 10. Camera coordinates can be mapped to the real world coordinates by the following transformation matrix $\begin{pmatrix} Xc \\ Yc \\ Zc \end{pmatrix} = Tcm \begin{pmatrix} Xa \\ Ya \\ Za \end{pmatrix}$, where Xc, Yc, Zc are the coordinates of the object in camera and Xa, Ya, Za are the coordinates of the same object in the real world and Tcm is the transformation matrix which can be calibrated for a camera.

7. The dimensions and inclinations of each line segment of the n × n segmented sub-square are the parameters used to represent an ideally flat terrain $Length (= Breadth)\ of\ each\ side = \frac{x}{n}\ Inclination\ of\ each\ side = 90^{\circ}$

8. Shortening of length (less than $\frac{x}{n}$) of any line segment (even skewed) of the formed lumigrids square mesh indicates that the terrain beneath the formed lumigrids is not flat. It is either concave or convex in nature along the Z axis.

9. The angle between the line segments (tangents of the line segments at the point of intersection if they are skewed) if not the right angle indicates that there is an inclination in XY plane of the terrain beneath the formed lumigrids based on the quadrant (first quadrant or fourth quadrant) of the inclination. Let a′ be the inclination of the line segments of the lumigrids and "a" the corresponding inclination in the ground is given by: $a = \pm d1 * a'$, where "d1" is the ratio of the inclination on the ground and the corresponding inclination caused by the lumigrids. And + indicates that the inclination is towards the first quadrant and—indicates that the inclination is towards the fourth quadrant.

10. After the image thresholding algorithm on the obtained image, the lumigrids are visible clearly as **Figure 7**.

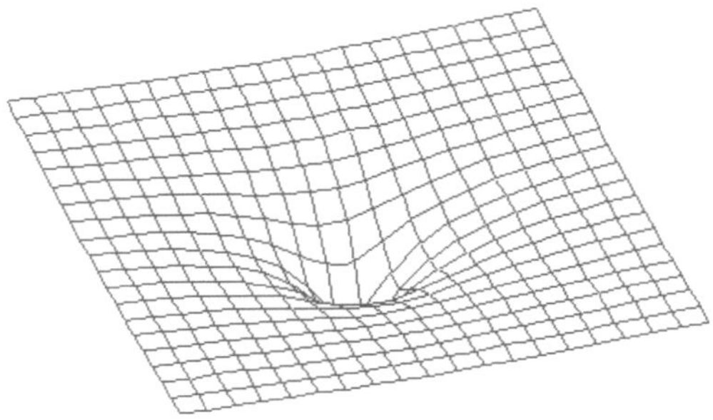

Figure 7.
Lumigrids formed over a pit.

In the above image, the required lengths between the skewed line segments are calculated.

11. Let a line segment of generated lumigrids of ideal expected size $\frac{x}{n}$ gets shorten by y% due to a skewed terrain. Let "d2" be the ratio of the absolute value of the vertical height on the ground indicated by the corresponding lumigrids to length of the corresponding line segment generated by the lumigrids. Then the absolute height "h" with reference to ideal flat surface of the ground is given by: $h = \pm d2 * \frac{x}{n} * \left(\frac{100-y}{100}\right)$. Axiom 5 decides if h is positive or negative. h is positive for concave terrain and negative for convex terrain. If y = 100%, theoretically there could be a narrow pit or hill in the ground, as indicated by the non-visibility of the lumigrids.

12. To exactly identify if the terrain at a given position is concave or convex in nature, we observe the inter line segment distance i of the terrain. If $i = \frac{x}{n} \rightarrow$ flat surface, if $i > \frac{x}{n} \rightarrow$ concave surface, if $i < \frac{x}{n} \rightarrow$ convex surface.

13. After calculating the terrain information of the given sub-square, the process is repeated to all the sub-squares so that the entire pedestrian path is scanned for its terrain details and mapped. The data thus obtained is pushed to the cloud.

The cloud now has precise information of the terrain. The pedestrian guidance phase consists of the following steps:

1. When the pedestrian wishes to navigate, the pedestrian's smart phone requests a route from source to destination. A GIS map is consulted to obtain various routes from the source to the destination. The data from the cloud has precise information about the terrain of each of the pedestrian paths presented in all these routes. An optimum route is selected based on the variations in the terrain in that route, pedestrian traffic density in the route, the route with easy help in case of danger or need and various other parameters which govern the safety of the pedestrian are considered.

2. The smart phone guides the pedestrian along this route in the pedestrian path. All major terrain variations in the pedestrian path are alerted to the pedestrian.

3. The shirt mounted unit on the pedestrian flashes the lumigrids on the path ahead and the camera embedded on the unit captures the image of the

Figure 8.
Lumigrids formed by the shirt mounted unit of a pedestrian in the guidance phase.

lumigrids formed and transmits this image to the smart phone of the pedestrian (**Figure 8**).

4. The terrain information obtained from the lumigrids are cross checked at real time with the terrain information available in the cloud to recognize and handle temporary terrain changes, like a dog sitting on the pedestrian's path or a random stone in the way, or sudden permanent terrain changes like a road block.

5. If considerable discrepancies are found in the terrain, the person is alerted to find possible alternate route like "Stop and Move 3 feet to your right" and a match for the known pattern in the cloud is checked for. If a match is found, the pedestrian is guided along that path.

6. If some permanent blocks are identified by the shirt mounted device, the cloud is notified about this so that the cloud can flag the terrain data of that pedestrian path as obsolete and can schedule a re-mapping of the terrain phase. An alternate route is found for the pedestrian and the pedestrian is guided accordingly.

Re-mapping of the terrain based on comparative walk-thru and terrain database phase consists of re-mapping of a pedestrian path either if the current data is flagged as obsolete by the pedestrian guidance phase, or a scheduled re-mapping process or on-need basis.

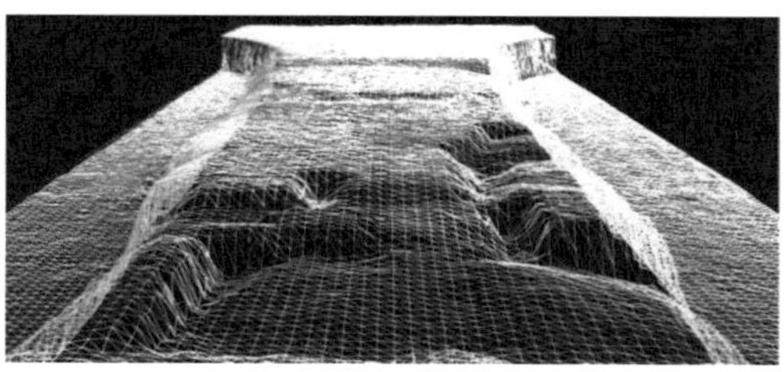

Figure 9.
Visualization of the terrain grid of a pedestrian path formed by the data.

The data on the cloud contains the terrain information of the pedestrian path capable of generating a terrain grid along with its GPS coordinates.

The visualization of the data represented as a terrain grid available on the cloud for a pedestrian path looks like **Figure 9**.

1. A sample data from the cloud is as follows

GPS	Ver.	h	a	Dirty bit
(20, 30)	1	+20	−3	0
(20, 31)	1	+20	−7	0
(20, 32)	1	+20	−10	0
(20, 33)	1	+20	−10	0
(21, 30)	1	+2	0	0
(21, 31)	1	+2	+1	0
(21, 32)	1	+3	0	0
(21, 33)	1	0	−2	0
(20, 30)	1	−8	0	0
(20, 31)	1	−8	0	0
(20, 32)	1	+2	0	0
(20, 33)	1	+2	0	0

GPS, the coordinates of the GPS location; Ver., the version number of the data; h, the height of the terrain; a, the inclination of the terrain; dirty bit, specifies if the data is obsolete

1. When the pedestrian wants to navigate, he first initiates a session with the cloud server which is a onetime activity for every navigation session.

2. The smart phone application now starts streaming the terrain data from the cloud shown above which is the reference data of the pedestrian path.

3. The system guides the person to follow the route and alerts on any terrain-related danger. For instance, when the pedestrian is in (20, 33). The interface alerts the pedestrian that there is a pit right in front of him ((20, 30), (20, 31) as indicated by a negative high value) and identifies that nearby terrain that is tolerable to walk and guides the pedestrian accordingly.

4. The lumigrids on the shirt scans the terrain ahead of the person and checks if there is an acceptable match with the reference data on cloud. If there is any discrepancy in the data obtained by the shirt and the cloud, the person is requested to take some alternative like a slight lateral movement and again a match is checked for. If the person is not able to get any help or no match is found, the server looks for alternative routes and guides the person. For instance, let the person be in (21, 30). According to the cloud data, there should be a high wall in front of him, but the shirt mounted unit scans and finds that there is no wall now and the terrain is optimum to walk. It flags all these data in the cloud as dirty by setting the Dirty Bit as follows:

GPS	Ver.	h	a	Dirty bit
(20, 30)	1	+20	−3	1
(20, 31)	1	+20	−7	1
(20, 32)	1	+20	−10	1
(20, 33)	1	+20	−10	1

Accordingly the cloud decides if it needs to schedule a re-mapping phase for that terrain or to accept the information shared by the pedestrian shirt.

After the re-map, following is the data in the cloud:

GPS	Ver.	h	a	Dirty bit
(20, 30)	2	+0	0	0
(20, 31)	2	+2	0	0
(20, 32)	2	+0	0	0
(20, 33)	2	+2	0	0

5.4 Summary

This section proposes a conceptual framework which fills the major gaps exist in the design of technological navigation aids and explains the software architecture, hardware and wearable devices requirements and the theoretical models necessary for building an infrastructure to seamlessly gather the terrain-related information of the pedestrian path and use this information to guide the pedestrians to navigate properly.

6. Conclusions

In this chapter, we outlined various aspects of wearable technology and its implementation in a wide range of applications, starting with healthcare, continued with other domains and concludes with the integration of wearables to navigation and safety systems. Wearables technology is still at its development and growing stage. We expect wearables to continue its fast growth and be implemented in much more domains, transforming our life to be much more convenient, safe and automated.

Author details

Menachem Domb
Computer Science Department, Ashkelon Academic College, Ashkelon, Israel

*Address all correspondence to: dombmnc@edu.aac.ac.il

References

[1] Kolodzey L, Grantcharov PD, Rivas H, Schijven MP, Grantcharov TP, Wearable Technology in Healthcare Society. Wearable technology in the operating room: A systematic review. BMJ Innovations. February 2017;**3**(1): 55-end

[2] Park S, Chung K, Jayaraman S. Wearables: Fundamentals, advancements, and a roadmap for the future. In: Wearable Sensors. Chapter 1.1. Academic Press, an imprint of Elsevier; 2014. pp. 1-23

[3] Liu Y, Wang H, Zhao W, Zhang M, Qin H, Xie Y. Flexible, stretchable sensors for wearable health monitoring: Sensing mechanisms, materials, fabrication strategies and features. Sensors. 2018;**18**:645. DOI: 10.3390/s18020645

[4] Alam MM, Hamida EB. Surveying wearable human assistive technology for life and safety critical applications: Standards, challenges and opportunities. Sensors. 2014;**14**(5):9153-9209. DOI: 10.3390/s140509153

[5] Armagan E, Papkovsky DB, Toncelli C. Chapter 2: New polymer-based sensor materials and fabrication technologies for large-scale applications. In: Quenched-Phosphorescence Detection of Molecular Oxygen: Applications in Life Sciences. Royal Society of Chemistry [RSC] publishing; 2018. pp. 19-49. ISBN: 978-1-78801-175-4

[6] Awolusi I, Marks E, Hallowell M. Wearable technology for personalized construction safety monitoring and trending: Review of applicable devices. Automation in Construction. 2018;**85**: 96-106

[7] Rodgers; Vinay MM, Pai; Richard M, Conroy S. Recent advances in wearable sensors for health monitoring. IEEE Sensors Journal. June 2015;**15**(6):3119-3126

[8] Majumder S, Mondal T, Deen MJ. Wearable sensors for remote health monitoring. Sensors. 2017;**17**:130. DOI: 10.3390/s17010130

[9] Mitchell R, Ozminkowski R, Serxner S. Improving employee productivity through improved health. Journal of Occupational and Environmental Medicine. 2013;**55**(10): 1142-1148. DOI: 10.1097/JOM.0b013e3182a50037

[10] Jin H, Jin Q, Jian J. Smart materials for wearable healthcare devices. In: Ortiz JH, editor. Wearable Technologies. Rijeka, Croatia: IntechOpen; 2018. DOI: 10.5772/intechopen.76604. Available from: https://www.intechopen.com/books/wearable-technologies/smart-materials-for-wearable-healthcare-devices

[11] Aliverti A. Wearable technology: Role in respiratory health and disease. Breathe. 2017;**13**(2):e27-e36

[12] Tambo E, Ngogang JY. Wearable nutrition and dietetics technology on health nutrition paradigm shift in low- and middle-income countries. International Journal of Nutrition and Metabolism. 2018;**10**(5):31-36. DOI: 10.5897/IJNAM2016.0207

[13] Javadi B, Calheiros RN, Matawie KM, Ginige A, Cook A. Smart Nutrition Monitoring System Using Heterogeneous Internet of Things Platform; 2017. Available from: staff.scem.uws.edu.au/~bjavadi/papers/javadi_IDCS2017.pdf

[14] Murakami H, Kawakami R, NakaeShow S, Miyachi M. Accuracy of wearable devices for estimating Total energy expenditure: Comparison with metabolic chamber and doubly Labeled water method. JAMA Internal Medicine. 2016;**176**:702-703. DOI: 10.1001/jamainternmed.2016.0152

[15] Lee J, Kim D, Ryoo H-Y, Shin B-S. Sustainable wearables: Wearable technology for enhancing the quality of human life. Sustainability. 2016;**8**(5): 466. DOI: 10.3390/su8050466

[16] Afzal A, Mousavi S-F, Khademi M. Estimation of leaf moisture content by measuring the capacitance. Journal of Agricultural Science and Technology. 2010;**12**:339-346

[17] Boukhechb M, Daros AR, Philip KF, Bethany C, Barnes TL. Demonic salmon: Monitoring mental health and social interactions of college students using smartphones. Smart Health. 2018;**9-10**: 192-203

[18] Alam MM, Arbia DB, Hamida EB. Research Trends in Multi-Standard Device-to-Device Communication in Wearable Wireless Networks, Part of the Lecture Notes of the Institute for Computer Sciences, Social Informatics and Telecommunications Engineering Book Series, LNICST; 2015. p. 156

[19] Banovic N, Franz RL, Truong KN, Mankoff J, Dey AK. Uncovering information needs for independent spatial learning for users who are visually impaired. In: Proc. ASSETS'13, Article: 24

[20] Sendero GPS for the Blind. Available from: www.senderogroup.com

[21] Trekker Breeze. Available from: www.humanware.com

[22] Lumigrids. Available from: http://www.gizmag.com/lumigrids-led-projector/27691/

[23] Kane SK, Jayant C, Wobbrock JO, Ladner RE. Freedom to roam: A study of mobile device adoption and accessibility for people with visual and motor disabilities. In: Assets'09. 2009. pp. 115-122

[24] Shoval S, Borenstein J, Koren Y. Mobile robot obstacle avoidance in a computerized travel for the blind. In: IEEE International Conference on Robotics and Automation. San Diego, CA; 1994